IJPHM

International Journal of Prognostics and Health Management

The International Journal of Prognostics and Health Management (IJPHM) is the premier online journal related to multidisciplinary research on Prognostics, Diagnostics, and System Health Management. IJPHM is the archival journal of the Prognostics and Health Management (PHM) Society. It exists to serve the following objectives:

- To provide a focal point for dissemination of peer-reviewed PHM knowledge.
- To promote multidisciplinary collaboration in PHM education and research.
- To encourage and assure establishment of professional standards for the practice of PHM.
- To improve the professional and academic standing of all those engaged in the practice of PHM.
- To encourage governmental and industrial support for research and educational programs that will improve the PHM process and practice.

The Journal supports these goals by providing a venue for archival publication of peer-reviewed results from research and development in the area of PHM. We define PHM as a system engineering discipline focused on assessing the current status and well as predicting the future condition of a component and/or system of components. PHM is broader than any single field of engineering: it draws from electrical, electronics, mechanical, civil, and chemical engineering, computer and materials science, reliability, test and measurement, artificial intelligence, physics, and economics. IJPHM seeks to publish multidisciplinary articles from industry, academia, and government in diverse application areas such as energy, aerospace, transportation, automotive, and industrial automation. IJPHM is dedicated to all aspects of PHM: technical, management, economic, and social.

IJPHM

International Journal of
Prognostics and Health Management

2012   Vol. 3 Issues 1-2

Table of Contents

http://phmsociety.org
Free and open access to full text papers worldwide.

Applying Weibull Distribution and Discriminant Function Techniques to Predict Damaged Cup Anemometers in the 2011 PHM Competition

Joshua Cassity[1], Christopher Aven[1], Danny Parker[1]

[1]Miltec Research and Technology, A Ducommun Company, Oxford, MS, 38655, United States
dparker@one.ducommun.com
jcassity@one.ducommun.com
caven@one.ducommun.com

ABSTRACT

Cup anemometers are frequently employed in the wind power industry for wind resource assessment at prospective wind farm sites. In this paper, we demonstrate a method for identifying faulty three cup anemometers. This method is applicable to cases where data is available from two or more anemometers at equal height and cases where data is available from anemometers at different heights. It is based on examining the Weibull parameters of the distribution generated from the difference between the anemometer's reported measurements and utilizing a discriminant function technique to separate out the data corresponding to bad cup anemometers. For anemometers at different heights, only data from the same height pair combinations are compared. In addition, various preprocessing techniques are discussed to improve performance of the algorithm. These include removing data that corresponds to poor wind directions for comparing the anemometers and removing data that corresponds to frozen anemometers. These methods are employed on the data from the PHM 2011 Data Competition with results presented.

1. INTRODUCTION

The issue of identifying faulty anemometers used during wind resource assessment at prospective wind farm sites has increased in recent years as wind energy grows in importance due to declining fossil fuel availability. In wind resource assessment, the need for effective wind estimation is critical. If anemometer readings differ from reality by a small amount, the cost in terms of return on investment can be large.

When data from anemometers at equal height above the ground (and therefore equal wind speeds in principle) are available, previous studies (Ye, Veeramachaneni, Yan., and Osadciw, 2009) have shown that the differences in wind speed between sensors at equal height can be characterized as a Weibull distribution. To identify broken anemometers the Weibull parameters of difference between wind speeds were estimated and then compared against thresholds for the shape and scale parameters. These thresholds where heuristically determined based on experience with previously good data. The researchers go on further to propose using the area under the cumulative distribution function (cdf) as a feature for discrimination. Though the method has demonstrated good results, there still remains the issue of analytically choosing an appropriate threshold. In this paper we utilize a discriminant analysis to generate the minimum-error-rate thresholds for distinguishing the bad sensor data sets from the good sensor data sets. The features we chose to discriminate on are the Weibull parameters characterizing the difference in the wind speeds between two anemometers. In cases where there were not anemometers paired at the same height a physics based model of the wind speed versus height was assumed to feed into the discriminant functions.

Discriminant analysis is a powerful set of tools that tries to analytically determine classification boundaries based on the statistical behavior of the features used to characterize the data. These boundaries can be further adjusted based on the probability of each of the classes occurring. Before the discriminant functions can be generated a certain amount of preprocessing must be done to condition the data. This is done to remove certain environmental and terrain effects that may skew the thresholds.

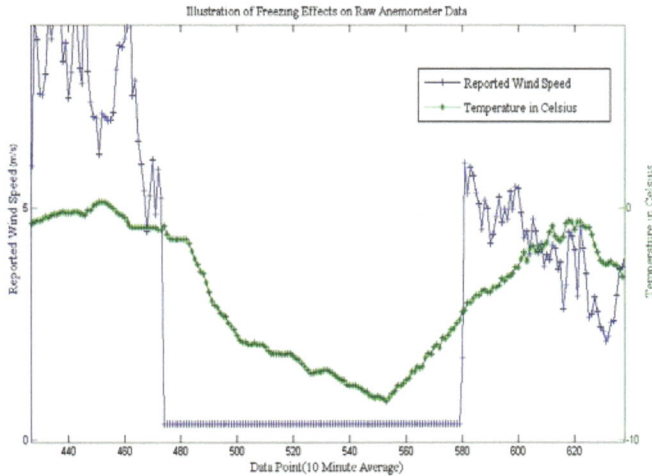

Figure 1-Effects of Freezing Conditions on Raw Anemometer Data

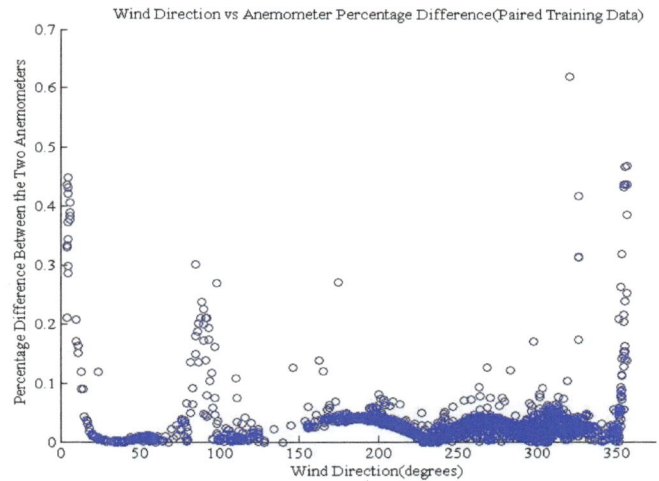

Figure 2-Wind Direction versus Reported Difference between Sensors at Equal Heights

The data from the PHM 2011 Data Competition was organized as follows. It was divided into two groups: 'paired' anemometer data containing two sensors at equal height and 'shear' anemometer data containing either three or four sensors at different heights above the ground. The wind speed measurements for each sensor were averaged over ten minutes and provided in the data, along with maximum, minimum, and standard deviation within each averaged ten minute segment. In addition, wind direction and temperature data were provided alongside the anemometer readings.

The goal of the PHM 2011 Data Competition was to determine, from provided data, whether given cup anemometers were damaged and reporting erroneous readings. Per competition rules, anemometers that become frozen due to weather effects do not count as damaged. These had to be identified in order to prevent false diagnoses. There are 420 test data files for the 'paired' case and 255 test data files for the 'shear' case. In the 'paired' case, a point is awarded when a data file is properly diagnosed, that is, when both sensors are correctly marked as damaged or undamaged. In the 'shear' case, a point was awarded if the competitor correctly determined that a sensor (if any) is damaged in the data file. In addition to the test data files, there were 12 training data files for both the 'paired' and 7 training data files for the 'shear' case. For these files, the anemometers guaranteed to be good, that is, not broken. So it was not necessary in the 'shear' case to identify the specific damaged anemometer. In the 2011 PHM Competition, submissions were graded and a leader board was provided to show the rankings of each team relative to one another. However, the actual scores for each submission were obscured and the labeled data was not released, making it impossible to accurately compare one algorithm's performance to another.

The rest of the paper is organized as follows. In 'Methodology' the data preprocessing, discriminant function technique, 'initial guess' estimation for paired data corresponding to anemometers at equal height, and 'initial guess' estimation for shear data corresponding to anemometers at different heights are presented, and in 'Conclusion' we mention some concluding remarks and point towards areas of future research. Throughout the paper, we will refer to data that corresponds to undamaged sensors as 'good data' and data from damaged sensors as 'bad data'.

2. METHODOLOGY

2.1. Data Preprocessing

There were several problems with the raw data used that had to be addressed before methods could be applied to determine faulty anemometers. At times, the anemometers would freeze and stop moving, even if they were not broken, and this would skew the algorithm towards overpredicting bad sensors. The method used to accommodate this was to search the data for measurements that were both below freezing and stuck on the lowest possible wind speed reported by the anemometer. This data was then discarded under the assumption that the sensor was frozen and reporting incorrectly.

For the 'shear' data set, if even one of the anemometers was seemingly frozen, all the data for that unit of time was discarded. Figure 1 gives an example of frozen data within the data set. One can clearly see where the reported wind speed drops to near zero which corresponds to temperatures below freezing.

A second issue that turned up was that, due to the anemometer placement (whether at 90 degrees to each other or 180 degrees), the observed mean difference between wind speed for anemometers at equal heights (such as found in the 'paired' data) would vary according to wind direction. This caused the differences to be more spread out because of the statistics in a few particular directions. To deal with this problem, each 'paired' data file was divided into bins 30 degrees in size. Then the standard deviation of the percentage differences between the two anemometers at each ten minute average was calculated for every bin and the two bins with the highest wind speed difference between the two anemometers were discarded. This threw out wind directions that corresponded to the greatest difference in wind speeds between anemometers at equal height, thus causing the remaining data to have a much lower standard deviation. In Figure 2, we show the variation between reported differences in wind speed versus wind direction for a training data set. In this figure data from 0 and 90 degrees would be discarded. If there were more than two directions that skewed the statistics they could also be discarded.

2.2. Discriminant Function Technique

The Discriminant analysis is a method of pattern classification that is known to achieve the minimum-error-rate classification for a given feature set (Duda, R. O, Hart, P. E., Stork, D. G., 2001). For each classification, a separate discriminant function is derived and evaluated for each feature vector. Then a particular set of data is determined to be of the class with the highest discriminant function value. This in effect yields the minimum error rate classification in the assigning of a class. For this application, there are two classes: the good sensors and the bad sensors and the features used to describe the class are the shape and scale parameters of the Weibull distribution of the difference in wind speeds. The formula for computing a discriminant function $g(\mathbf{x})$ for arbitrary covariance matrix is given by,

$$g_i(\mathbf{x}) = \mathbf{x}^t \mathbf{W}_i \mathbf{x} + \mathbf{w}_i^t \mathbf{x} + w_{i0} \qquad (1)$$

Where,

$$\mathbf{W}_i = -\frac{1}{2} \mathbf{\Sigma}_i^{-1} \qquad (2)$$

$$\mathbf{w}_i = \mathbf{\Sigma}_i^{-1} \boldsymbol{\mu}_i \qquad (3)$$

And,

$$w_{i0} = -\frac{1}{2} \boldsymbol{\mu}_i^t \mathbf{\Sigma}_i^{-1} \boldsymbol{\mu}_i - \frac{1}{2} ln|\mathbf{\Sigma}_i| + ln\, P(\omega_i) \qquad (4)$$

Here $\mathbf{\Sigma}$ and $\boldsymbol{\mu}$ refer to the covariance matrix and mean vectors of the features, respectively. The mean vector can be understood as the center of a class and the covariance matrix describes how scattered the points in the class are distributed about the center. $P(\omega_i)$ refers to the a priori probability that a given test file belongs to the class which

the discriminant function corresponds to. In equation 1, $\mathbf{x}$ refers to the feature vector for a particular set of data under consideration, that is, the Weibull parameters generated from the percentage difference between the two anemometers. This number could typically be obtained from the failure rate of the sensors. Since that was not known it was estimated by examining the percentage of the files that the initial estimation classified as bad. To apply the Discriminant Function Technique, it is necessary to have labeled or known data from both classes. Since there was no labeled bad data provided, a method had to be devised to get an initial guess at the 'bad sensor' class. Exactly how the initial classifications were determined for both the paired and shear cases is described in the next section. If more detailed information on discriminant function analysis is desired (Wolverton, C., Wagner, T., 1969) is an excellent resource.

2.3. Paired Data

Previous research (Ye et al, 2009) has demonstrated that anemometers at similar heights exhibit a Weibull distribution in the time domain differences between their mean reported wind speeds. In their paper, the shape and scale parameters of the Weibull distribution from a week's worth of data is estimated using a maximum likelihood estimation and results over a number of weeks are plotted on a graph with the scale parameter on the x-axis and shape parameter on the y-axis. Then, a visual investigation is performed and a 'cloud' of good performing sensors is identified by drawing an oval around the area of highest density, leaving the points outside the oval to be flagged as bad. This analysis was applied to obtain the initial bad set of files corresponding to good sensors. In our case, since the data in the PHM 2011 Data Competition was provided in sets of five days, we used five days as the interval of time for which to calculate scale and shape parameters. From this set of data, we employed two methods to label files as either good sensor or bad sensors for the discriminant analysis. Secondly, a hypothesis test was used to determine if a set of data came from a Weibull distribution from the good sensor class. The Kolmogorov–Smirnov (K-S) test (Eadie, W. T., Drijard, D., James, F. E, Roos, M., & Sadoulet, B., 1971) was implemented as the hypothesis test with the null hypothesis being that the data under consideration was from the same distribution as a good pair of sensors.

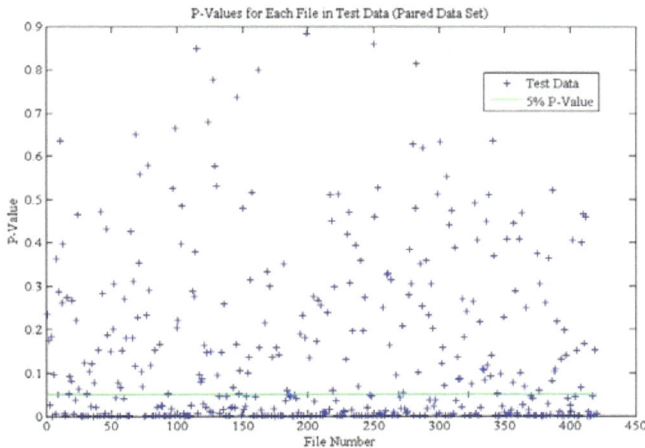

Figure 3-Confidence Level for each Test File in rejecting the null hypothesis

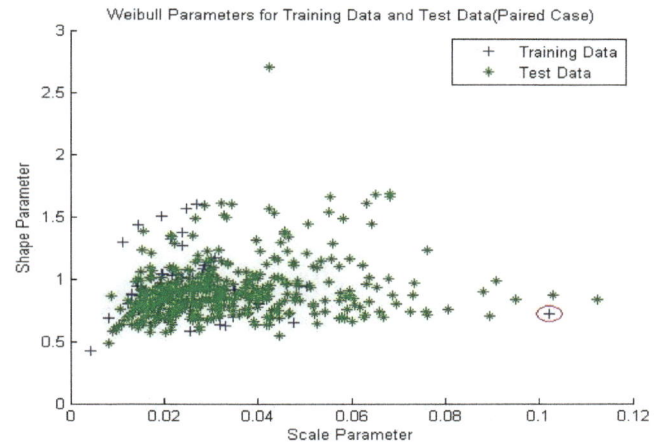

Figure 4-Scatter Plot of Weibull Parameters for all Training and Test Files in Paired Data

To accomplish this, the percentage difference between the anemometers was calculated for every 10 minute average in each of the training files (which were known to be good data). Each training file formed a distribution of percentage differences. Then, for each of the test files, the same process was undertaken and the K-S test was performed for the test file against each one of the 'good' distributions. If the null hypothesis, that the test file and training file were from the same distribution, was rejected for all 'good' distributions with 5% significance level, then the file was flagged as bad. On the other hand, if the test file's distribution could be matched with at least one training file at 5% significance level, then the test file was marked good. In order to be assured that the initial data in the bad class was truly from bad cases the intersection of the K-S set and the set derived from visual inspection of the data (that is, the files common to both sets) was used to obtain the set of bad sensors for applying the discriminant analysis.

Figure 3 shows the confidence level of rejecting the null hypothesis for each test file, given the set of training data files. As can be seen, this method is another form of thresholding where the threshold in this case is the significance level. A significance level below 5% indicates that there is a less than five percent probability that the null hypothesis is correct.

Figure 4 shows the plot of the Weibull parameters for the paired data set, both training and test data. Note the 'cloud' of data points in the lower left corner clustered together corresponding to the estimated 'good' data files. So in general 'good' data files have lower scale parameters than 'bad' data files. A single training data point seems disconnected from the rest of the training point, with a much higher than expected scale parameter. This point (circled in the figure) was discarded when choosing our threshold, as it corresponded to a data file where a large amount of points had been thrown out due to the wind direction preprocessing described earlier.

For the paired data, a competition requirement was to identify not only when a sensor was bad but which one of the pair was defective. In order to identify which specific anemometer has failed, the assumption was made that the anemometer with the lower mean reported wind speed will be the one that is bad. This was based on a consideration of the types of damage possible to cup anemometers. For example, a chipped or cracked cup will not hold wind as well as a normal one and as such should report a lower wind speed.

2.4. Shear Data

In the shear data, if there were sensors paired at each height then the previous methods for getting initial labeled data would be effective. However, that was not the case here. Data were given over various days with only one sensor at each height and to make matters more difficult, the heights were not consistent from file to file.

So to generate some initial labels for both good and bad classes, a physics-based model was employed. There are two different models that can be used to characterize the relationship between wind speed and height. The first, referred to as the wind profile power law (Oke, T, 1987). It is of the form,

$$\frac{u}{u_r} = \left(\frac{z}{z_r}\right)^\alpha \qquad (5)$$

Where u and z is the mean wind speed and height, respectively under consideration, u_r and z_r refer to the wind speed and height at a given reference point (usually 10 meters), and α is an empirically derived constant whose value depends on the stability of the atmosphere. For conditions of neutral stability, α is approximately 0.143. This equation assumes the relationship follows a simple power relation (Touma, J, S, 1977).

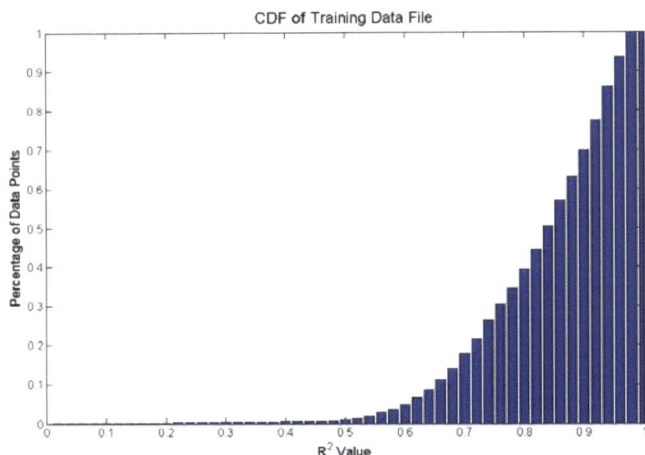

Figure 5-CDF of R^2 for a training file in the Shear data set

This equation does not take into account certain terrain features such as the roughness of the surface or the level of atmospheric stability, which can greatly affect the reported wind speeds. It only requires the mean wind speed at a 'reference' point, usually 10 meters. The second model, referred to as the 'logarithmic wind profile law' (Oke, 1987), is shown in equation 6,

$$u_z = \frac{u_*}{\kappa}\left[ln\left(\frac{z-d}{z_0}\right) + \psi(z, z_0, L)\right] \quad (6)$$

Equation 6 is valid from the surface up to around 1000m. It takes stability and surface roughness into account, but requires a number of known parameters, such as the zero plane displacement, friction velocity, and the Monin-Obukhov stability parameter, none of which were available to us.

Since the physical parameters were not known, a general *log*-linear relationship of the wind speed versus height from the ground was assumed. With the assumption that the data should fit a logarithmic curve, the exponential of the data was taken. The results of which should produce wind speeds that are a linear function of height. Once this conversion has been made, a simple linear regression analysis was performed to determine residuals and goodness of fit. The result of the linear fit analysis, R^2, gives a quantifiable value that can be used to classify on a sample by sample basis. However, the problem still remained on what an appropriate R^2 threshold should be to label a file as bad. To attempt at a systematic way to arrive at a threshold the R^2 values were then used to make histograms for estimating the probability density function (PDF), with careful consideration given to keeping bin edges equal. The PDF's were then integrated to form the cumulative density function (CDF). The CDFs represent the percentage of files equal to or below a particular R^2 value.

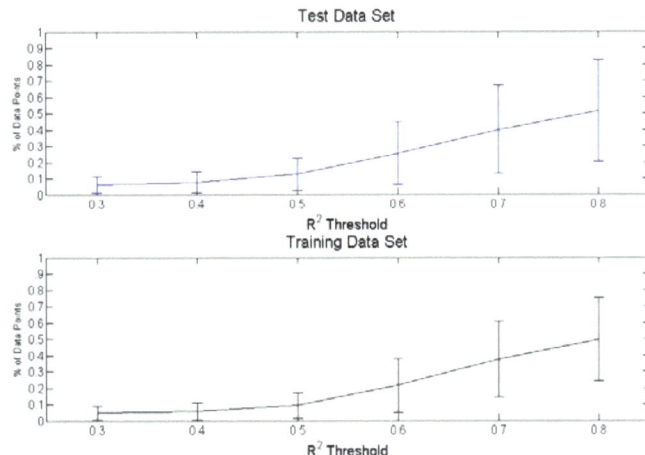

Figure 6-Mean and Standard Deviation for each R^2 value analyzed in Shear data

Figure 5 shows a CDF plot for a typical training data file. Notice that there are very few files that don't have at least a R^2 value above 0.7. This is in line with expectations that a *log*-linear fit does a good job modeling the wind profile as a function of height.

To arrive at an exact threshold value the CDF's were then analyzed at six values of R^2, ranging from 0.3 to 0.8, to determine what percentage of data points per file were less than or equal to the R^2 values of interest. The idea being that the training data would have less percentage of data points below a particular R^2 value when compared to test data that has potential bad sensors in it. At each threshold the mean and standard deviation of the number of files labeled as bad were then calculated for both training and test data. Figure 6 shows the plots of mean and standard deviation for each R^2 value analyzed.

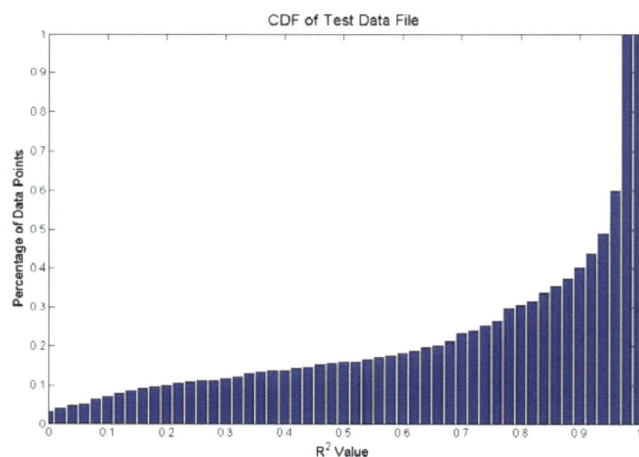

Figure 7-CDF of R^2 for a test file in the Shear data set

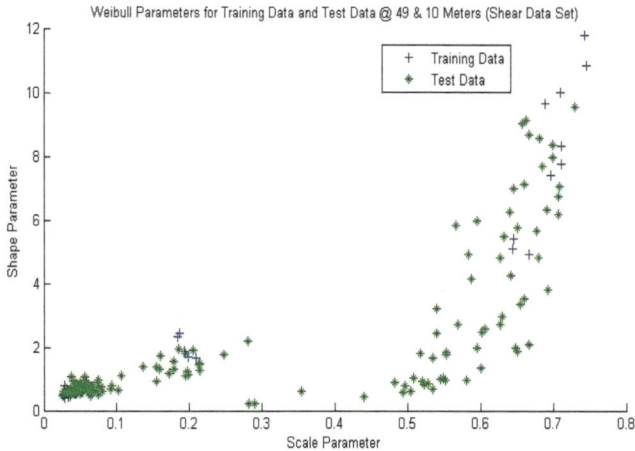

Figure 8-Scatter Plot of Weibull Parameters for all Training and Test Files in Shear Data (49 & 10 Meter Case)

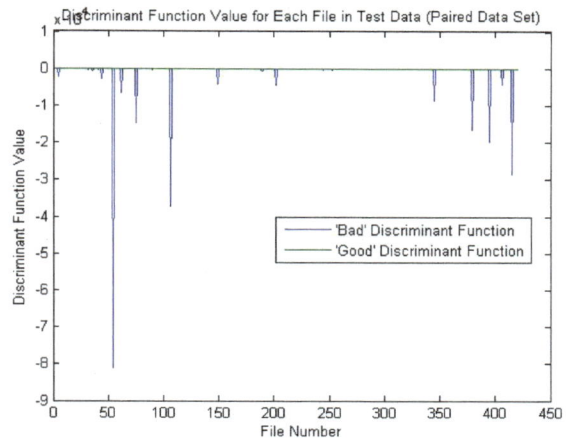

Figure 9-Discriminant value comparison for each test file in Paired Data Set

The R^2 value with the largest discrepancy between training and test data was then selected as the threshold value for labeling a file as bad. The assumption was made that the larger difference between the two data sets at a given R^2 value would indicate that the optimal threshold for discriminating between the two classes was achieved.

Since this threshold is on a per sample basis and the desire of the competition was to label the data per day, it had to be worked out how many files flagged during a day would cause the data to be labeled as bad. To come up with this threshold the per sample R^2 threshold was analyzed to determine the max percentage of training data points that were included in all bins less than or equal to the chosen threshold. Since it was known that all training data was considered good then it was assumed that whatever percentage of training files labeled as bad represented some acceptable percentage per day. Any days that had a higher percentage would then be labeled as bad. In practice a value slightly higher than this max value was then chosen as the percentage threshold for damage detection. This slight increase in percentage threshold value was selected to ensure that no training data was flagged as bad. The CDFs generated from every test file were then evaluated at the R^2 per sample threshold. Figure 7 shows the CDF of a typical test data file. The percentage of data points contained in all bins less than or equal to the per sample threshold was then compared to the percentage threshold obtained from the training data. If the percentage of test data points contained in those bins were greater than or equal than that threshold, the file was marked as bad.

Using this line of reasoning to label some files as the bad sensors class, we are able to extend the previously discussed discriminant analysis to the shear data.

However there is still some difficulty in using this approach for sensors at different heights. The problem is that the percentage difference between anemometers will increase or decrease with different height differences. To help alleviate this issue, all the training data was divided into height-pair combinations (such as 49 & 10 meters, 35 & 10 meters, etc), Weibull parameters were generated for the percentage difference between the respective sensors, and the discriminant function method was employed on these parameters using the files labeled earlier as the 'bad set'. This allowed us to compare each sensor to every other sensor in a data set. If a sensor goes bad, ostensibly this will show up in the Weibull parameters generated by the comparison of that sensor with all other sensor in the data set.

2.5. Discriminant Analysis Summary

In this section, the discriminant analysis employed in this paper will be briefly summarized. The analysis relies on creating discriminant functions for each desired classification ('good' and 'bad'). The equations to create this function are given in equations (1)-(4). The necessary parameters are the feature vector, in this case, the scale and shape parameters derived from the Weibull distribution of the percentage difference between two sensors under test. The resulting feature vector is of dimension 2 x 1. By examining all the feature vectors for both groups of labeled data yields two mean vector of size 2 x 1 and two covariance matrices of size 2 x 2, one for each class. In order to obtain the data necessary for each discriminant function, the 'training data' provided by the competition was used as good data and a variety of methods outlined in the previous two sections were used to obtain an estimate of some initial good data. Once the feature vectors, mean vectors, and covariance matrices were obtained for both classes, the parameters of a Weibull distribution describing the difference between mean wind speeds of sensors at

6

different heights were estimated. The resulting shape and scale parameters were used as the vector x in both discriminant functions, good sensors and bad sensors. The file was labeled as either good or bad based on which discriminant function produces the highest value. This process was repeated for every file under consideration.

After the discriminant functions have been created, they are evaluated at each file and the file is classified into the group whose discriminant function has the higher reported value. In our analysis, using the 'initial guess' presented in the previous two sections, we flagged 25% of the files as bad from the 'paired' data set and 58% of the files from the 'shear' data set. Figure 9 illustrates the outputs of the discriminant functions for each test file in the paired data. The function with the higher value will correspond to the minimum error rate classification of incorrect classification.

3. CONCLUSION

Building on the previous work of others in identifying that a Weibull distribution can statistically describe the differences between paired anemometers over short distances, we have proposed a conceptually simple method using discriminant functions for analytically determining classification thresholds. There are several complicating parameters like not having paired data at all heights and consistent heights, that are most likely artifacts of the competition and not indicative of real world monitoring. In addition, a real world application would generally also provide environmental information such as stability & surface roughness, along with the Monin-Obukhov stability parameter, which would enable the more accurate 'logarithmic wind profile law to be used. The performance of the data was improved by preprocessing to remove obviously faulty data and there was a rough attempt at estimating the probability of a sensor being bad. There are several ways in which this method could be improved in the future. One improvement could be had if the statistics between all sensors on a tower were modeled and used as features. This would allow for a more complete and robust description of an installation which in turn would allow for more powerful classification techniques to be applied.

Though the discriminant function yields the minimum-error-rate classification, it is highly sensitive to variations in the intial guess. We noticed that varying the files in the initial guess can dramatically alter the classifications. Therefore, finding the best method to obtain the initial guess of the bad files is critical.

Also, a better method for accommodating the variation in wind speed differences between anemometers at the same height with wind direction would offer some improvement. A simple method of doing so would be to find a function that characterizes the plot of wind speed differences versus wind direction as presented earlier and subtract the effect from the data. To accomplish this, it would be necessary to know the precise orientation of the anemometers ahead of time as this is not always possible to deduce from the plot of the data (sometimes wind may be from a small number of directions for the duration of a test and plot such as Figure 2 cannot be easily made). Another possible improvement would be if additional data such as surface roughness, and atmospheric stability information is available, then the log wind profile equation can be used in which should greatly improve the predictive ability for the shear data set.

REFERENCES

Duda, R. O, Hart, P. E., Stork, D. G., (2001). *Pattern Classification.* New York, NY: John Wiley & Sons, Inc

Eadie, W. T., Drijard, D., James, F. E, Roos, M., & Sadoulet, B. (1971). *Statistical Methods in Experimental Physics.* Amsterdam, North-Holland: Elsevier Science Ltd

Oke, T, R, (1987). *Boundary Layer Climates*, London, UK: Routledge

Touma, J, S, (1977). Dependence of the wind profile power law on stability for various locations, *J. Air Pollution Control Association*, Vol. 22, pp. 863-866.

Wolverton, C., Wagner, T. (1969), Asymptotically optimal discriminant functions for pattern classifiers, *IEEE Transactions on Information Theory*,

Ye, X., Veeramachaneni, K., Yan, Y., & Osadciw, L. A. (2009), Unsupervised Learning and Fusion for Failure Detection in Wind Turbines, *Proceedings of 12th International Conference on Information Fusion.* July 6-9,Seattle,WA

Feature Extraction and Pattern Identification for Anemometer Condition Diagnosis

Longji Sun[1], Chao Chen[2], and Qi Cheng[3]

[1,2,3] *School of Electrical & Computer Engineering, Oklahoma State University, Stillwater, OK, 74078, USA*
longji.sun@okstate.edu
chao.chen@okstate.edu
qi.cheng@okstate.edu

ABSTRACT

Cup anemometers are commonly used for wind speed measurement in the wind industry. Anemometer malfunctions lead to excessive errors in measurement and directly influence the wind energy development for a proposed wind farm site. In the PHM 2011 Data Challenge Competition, two types of data need to be processed for anemometer condition diagnosis: paired data consisting of wind data from paired anemometers, and shear data composed of measurements from an array of anemometers at different heights. Since the accuracy of anemometers can be severely affected by the environmental factors such as icing and the tubular tower itself, in order to distinguish the cause due to anemometer failures from these factors, our methodologies start with eliminating irregular data (outliers) under the influence of environmental factors. For paired data, the relation between the normalized wind speed difference and the wind direction is extracted as an important feature to reflect normal or abnormal behaviors of paired anemometers. Decisions regarding the condition of paired anemometers are made by comparing the features extracted from training and test data. For shear data, a power law model is fitted using the preprocessed and normalized data, and the sum of the squared residuals (SSR) is used to measure the health of an array of anemometers. Decisions are made by comparing the SSRs of training and test data. The performance of our proposed methods is evaluated through the competition website. As a final result, our team ranked the second place overall in both student and professional categories in this competition.

1. INTRODUCTION

Wind energy as a promising renewable energy source has attracted considerable attention in recent years. The first step in the development of a productive wind farm is wind re-source assessment. Cup anemometers (IEA, 1999) have been widely used for wind speed measurement. Typical anemometers have three or four cups installed on a vertical shaft. Their measurements provide important information of wind resources for a proposed site. Therefore, their accuracy can greatly affect the estimated energy production and return on investment. Normally, the measurement of a cup anemometer is within 2% error. However, under some circumstances, such as the wear on the bearings, a missing cup or a failed shaft, an anemometer fails to provide accurate wind speed information, i.e., its measurements have excessive errors. It is critical that damaged or out of tolerance anemometers be detected and replaced in a timely manner.

Recent years have seen various methods proposed for the anemometer condition diagnosis problem. In (Beltran, Llombart, & Guerrero, 2009b), the nacelle anemometer fault detection problem is studied, in which wind speeds at one target anemometer are estimated by using two reference anemometers in its vicinity and the deviations of the estimates from the measurements are used to determine the target anemometer's condition. In (Beltran, Llombart, & Guerrero, 2009a), a method is introduced to select the range of data so that the uncertainty in evaluation of anemometers' health is minimized. To predict the failure of a hot-wire anemometer, a method utilizing a feature related sensor degradation and analyzing the trend of the feature is proposed (Delfino, Puttini, & Galvao, 2010). In the work by Kusiak, Zheng, and Zhang (2011), a virtual speed sensor is built based on historical wind speed data to monitor real sensors. In (Siegel & Lee, 2011), an anemometer assessment methodology using residual processing and clustering techniques is proposed, in which the residuals of anemometers' readings are computed and clustered to determine the anemometers' conditions.

The PHM 2011 Data Challenge is focused on the detection of failed anemometers. Generally, anemometers are installed on a meteorological tower. With single or paired anemometers at different heights, an array of anemometers is formed.

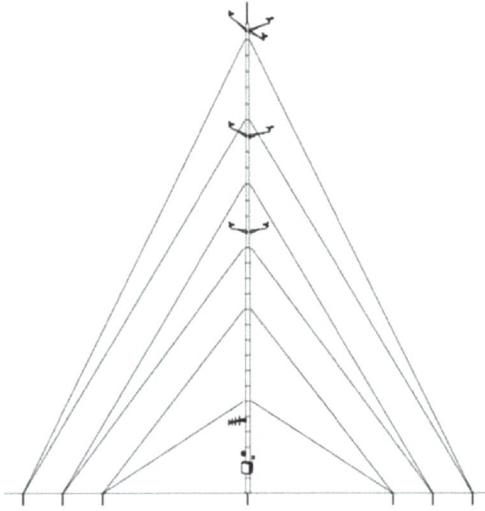

Figure 1. An example 60m tower with three sensor locations (https://www.phmsociety.org/competition/phm/11/problem).

Figure 2. Normalized positions of sensors on a tower (https://www.phmsociety.org/competition/phm/11/problem).

Figure 1 shows an example of a 60m meteorological tower with sensors located at 59m, 49m and 39m, respectively. Figure 2 shows the normalized position of sensors at a tower. A paired data file includes the measurements of paired anemometers with a $90°$ or $180°$ angle, the corresponding wind direction and temperature. Each shear data file includes the measurements of an array of anemometers, wind direction, temperature and data collection time. The paired data consist of 12 training sets and the shear data consist of 7 training sets, each consists of 25 days of normal data. The paired data also have 420 test files and the shear data 255 test files, each of 5 days of data of unknown conditions. The problem is to detect failed anemometers in each test file. For paired data, it is to distinguish which one, or both of anemometers, fail if not both of them work normally. The objective for each shear test file is to determine whether all anemometers are in a good condition or not. Readers are referred to the PHM 2011 Data Challenge website (https://www.phmsociety.org/competition/phm/11/problem) for more information.

Since, in training files, only normal data are provided, the problem of anemometer fault detection is essentially anomaly detection. Various techniques have been developed for anomaly detection, including classification based methods (Duda, Hart, & Stork, 2000), statistical approaches (Barnett & Lewis, 1994), and clustering techniques (R. Smith, Bivens, Embrechts, Palagiri, & Szymanski, 2002). Anomaly can be categorized into point anomaly, contextual anomaly, and collective anomaly (Chandola, Banerjee, & Kumar, 2009). Point anomaly, i.e., anomalous individual data instance, is the most studied anomaly and the focus of most of the existing anomaly detection techniques. Contextual anomaly

refers to a data instance only considered as an anomaly in a specific context. For example, in the work by Basu and Meckesheimer (2007), anomalies in time series data are detected by comparing the value of a data point with the median of its neighborhood. Collective anomaly means that a collection of data instances is anomalous, in which the relation between data is exploited to detect anomalies. For instance, sequential anomaly detection techniques are used to find unusual values in multiple time-series data (Chan & Mahoney, 2005).

In this paper, we will extract from training data important features that can reflect normal collective patterns or behaviors of anemometers in various contexts. Any deviation from these normal patterns can indicate possible faulty conditions. The rest of the paper is organized as follows. In Section 2, the methodology to analyze the paired dataset is provided. The method to deal with shear data is elaborated in Section 3. The paper is concluded with some discussion in Section 4.

2. METHODOLOGY FOR PAIRED DATA ANALYSIS

The method for paired data analysis mainly includes five steps: data preprocessing, feature extraction, denoising, pattern search and decision making. Firstly, a preprocessing step is taken to eliminate some apparently incorrect and statistically useless measurements. Secondly, a feature, namely, the relation between the discrepancy of the paired anemometer measurements and the wind direction, is extracted from the preprocessed data. A further denoising step is taken to reduce the environmental effects and make the feature more prominent in different situations. Then, an algorithm is designed to search for each test data file the most matched pattern from training data. Finally, decisions are made based on the relation between the pattern under testing and the matched pattern.

2.1. Data Preprocessing

Failed anemometers cannot provide accurate wind speed measurements. On the other hand, environmental factors, such as icing can also affect the accuracy of measurements considerably. To avoid false alarms, it is important to distinguish these two types of situations. Some preprocessing of the raw measurements is required.

The preprocessing step is composed of two stages. In the first stage, data undergo a measurement range test. Namely, only measurements within a valid measurement range are meaningful. Factors, such as sensor noise and icing, result in measurements outside this range, which fail to provide useful information and should be eliminated. For this problem, the range is set to be from 0.4m/s to 75m/s (https://www.phmsociety.org/competition/phm/11/problem).

In the second stage, detection of icing conditions is conducted. Icing is a leading factor in introducing errors in measurement data. Empirical results (Kenyon & Blittersdorf, 1996) and our observations of the training data have shown that icing conditions have the following characteristics:

1) When the temperature is at or below the icing point, the standard deviation of the wind speed measurements is zero or near zero.

2) The standard deviation of the wind direction measurements is zero or near zero.

In (Schaffner, 2002), it is suggested that the measurements in six hours before and after the icing points should be discarded, considering that the effect of icing begins long before an anemometer is frozen and continues for some time before the frozen effect completely disappears. Since we have limited data in this competition, especially for test data, a more practical range is adopted in which only the data in 30 minutes before and after icing points are discarded.

2.2. Feature Extraction

In an ideal environment, the measurements of a pair of normal anemometers should be very close to each other given that they measure the wind speeds at the same height with a very close distance. However, this is not always the case for the given training data. It can be shown that the mast of the tubular tower on which the paired anemometers are mounted plays an important role (Lubitz, 2009). The mast of the tower will generate a wake behind it, acceleration around it and a retardation upwind of it (IEA, 1999). Figure 3 shows a wind field around the mast of a tubular tower. The numbers indicate the ratio between local wind speeds and the free-field wind speed. This fact explains the significant difference in

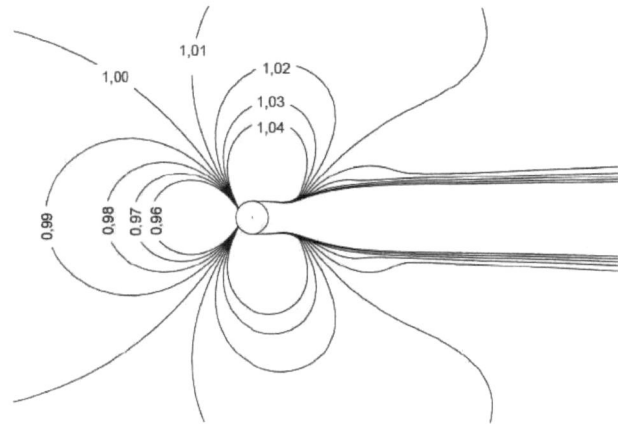

Figure 3. The wind field around a tubular mast (from the IEA 1999 report).

paired anemometer measurements in some wind directions. This suggests that the relation between the wind speed difference and its corresponding wind direction can be utilized as a key feature to describe the condition of paired anemometers. The wind speed difference is computed as follows:

$$s = \frac{s^{(1)} - s^{(2)}}{\max(s^{(1)}, s^{(2)})} \tag{1}$$

where $s^{(1)}$ is the wind speed of anemometer 1 and $s^{(2)}$ is the wind speed of anemometer 2. Normalization is taken to simplify the subsequent pattern search step. This is different from (Lubitz, 2009), where $s = s^{(1)}/s^{(2)}$ is used as a wind difference indicator to evaluate the tower effect approximation model. Figures 4(a) and 4(b) show the normalized wind speed difference as a function of the wind direction for pairTrng1 and pairTrng7 for example. Figures 5(a) and 5(b) plot the same relation for two test data files. Since the training data are from normal anemometers, the relations between the wind speed difference and the wind direction based on these training files are the representatives of normal behaviors of anemometers. Deviations from these representative patterns may indicate failure of anemometers in test data.

2.3. Denoising

In Figure 4(a), we observe that around 90° and 360°, the wind speed difference deviates from zero, while for the rest of wind directions, the difference varies around zero. This may be due to the normalized position of the paired anemometers with respect to the mast. Besides, there are some data points isolated from the majority of the rest, which are marked with circle in the figure. This situation is more severe in test data. Because of the limited size, the percentage of isolated points in test data can be large. By checking the original data, the isolated data points generally correspond to a low wind temperature when anemometers may run slow. This is the case for all training data. To make the pattern more prominent and make

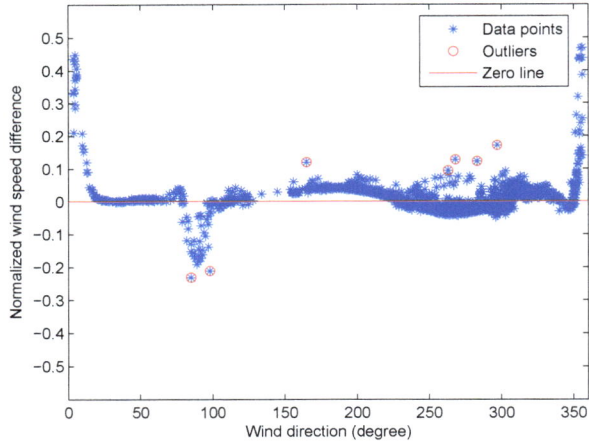

(a) PairTrng1

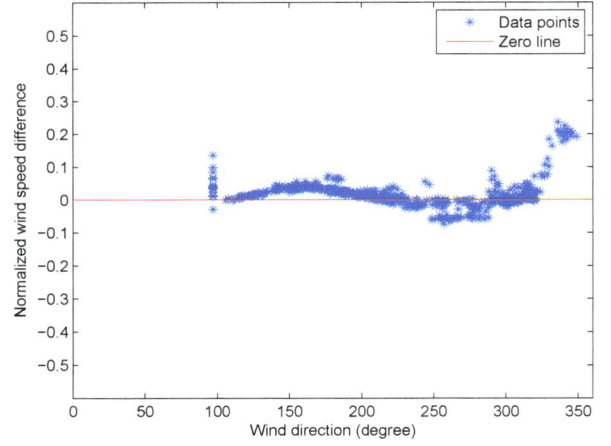

(b) PairTrng7

Figure 4. Normalized wind speed difference as a function of wind direction for training data.

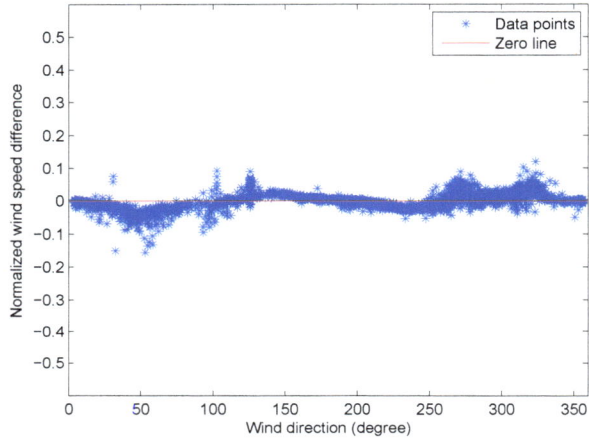

(a) Pairdata1

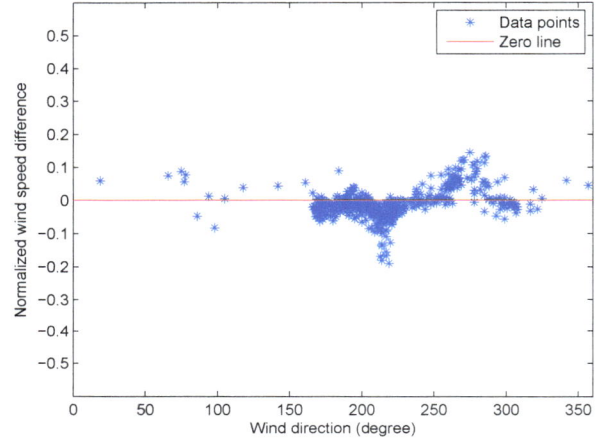

(b) Pairdata12

Figure 5. Normalized wind speed difference as a function of wind direction for test data.

it easier for test data to find a match, we consider these isolated points as outliers and they should be eliminated.

There are many ways to remove outliers. The method adopted here is based on the average distance of each data point from its k nearest neighbors, the larger of which indicates it is more likely to be an outlier. More information about the distance-based outlier detection techniques can be found in (Knorr, Ng, & Tucakov, 2000). The distance between two data points in Figure 4(a) is defined as follows:

$$D = \sqrt{\left(\frac{d_i - d_j}{360}\right)^2 + (s_i - s_j)^2} \qquad (2)$$

where d_i (d_j) is the wind direction and s_i (s_j) is the wind speed difference along that direction. Normalizing d_i by 360° is to make these two quantities comparable. For every data

point (d, s), the average distance from its k nearest neighbors is calculated,

$$D_{(d,s)} = \frac{1}{k}\sum_{i=1}^{k}\sqrt{\left(\frac{d - d_i}{360}\right)^2 + (s - s_i)^2} \qquad (3)$$

where $\{(d_i, s_i), i = 1, \cdots, k\}$ is the set of k nearest neighbors. Since the distribution of data points is different at different wind direction, we compare $D_{(d,s)}$ only with that of those data points of similar wind directions. A window of length Δd moves along the wind direction axis. For all data points in this window, those whose average distance is among the largest $\alpha\%$ are marked as outliers and are eliminated. The performance of this method depends on parameter k, Δd and α. In our experiments, we set $k = 10$, $\Delta d = 20$ and $\alpha = 10$ which gives good empirical results. Figures 6(a) and 6(b)

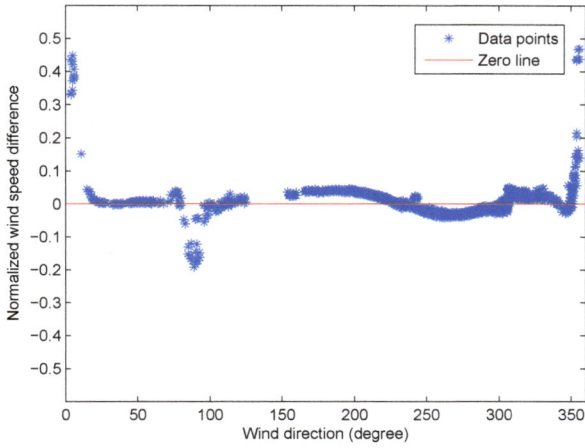

(a) PairTrng1

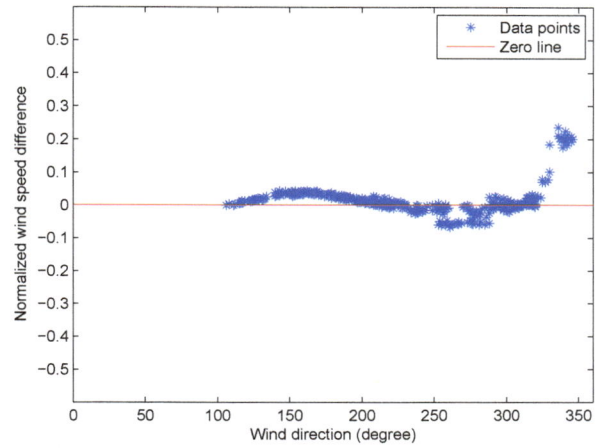

(b) PairTrng7

Figure 6. Normalized wind speed difference vs wind direction for training data after denoising.

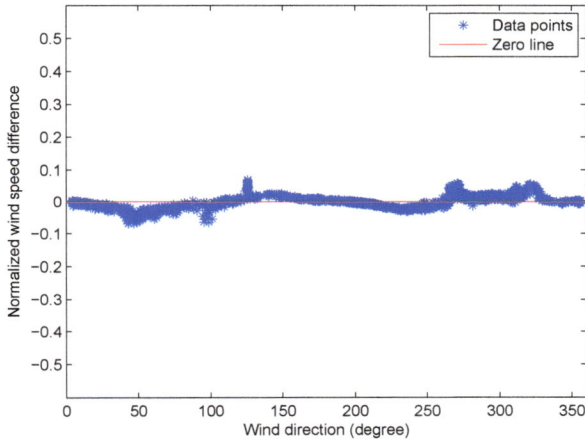

(a) Pairdata1

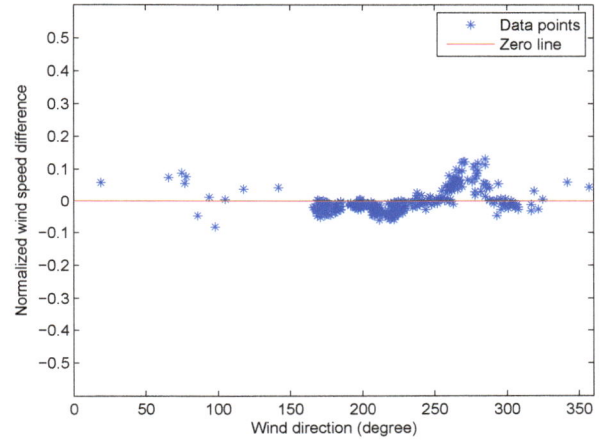

(b) Pairdata12

Figure 7. Normalized wind speed difference vs wind direction for test data after denoising.

are the relation of wind speed difference and wind direction after removing outliers (isolated points). This can also be applied to test data files and the results are shown in Figures 7(a) and 7(b).

2.4. Pattern Search

Training data are collected from normal anemometers. Since there are twelve training files, there are twelve normal patterns under different configurations. In this step, we need to find, for each test file, the most matched training profile for comparison. Distance is the most used metric to measure the similarity of two patterns. Given a training file p and a test file q, assume that there are a total of N wind directions that both training and test files have wind speed difference values. If these values are plotted in an N-dimensional space,

two point clouds are formed. Figure 8 is an example when $N = 2$. The distance of two point clouds can be measured by the distance between their centroids. More specifically,

$$Dis(q,p) = \frac{1}{N}\|\overline{S}_q - \overline{S}_p\|_2 \qquad (4)$$

where $\overline{S}_q$ is an N-dimensional vector, each element of which is the mean wind speed difference for that wind direction. The same applies for $\overline{S}_p$, $p = 1, \cdots, 12$. Normalization over N is done to eliminate the effect of the number of dimensions. Another important factor is the shape of data distribution pattern. The similar shape indicates a similar anemometer configuration. The correlation coefficient is adopted and defined

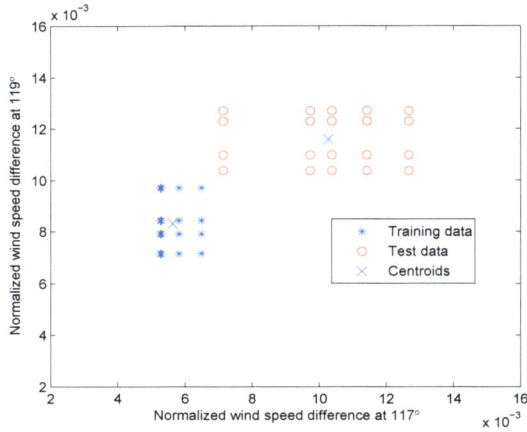

Figure 8. Two point clouds for training data (∗) and test data (◦) when wind directions 117° and 119° are selected. "×" represent the centroids of the clouds.

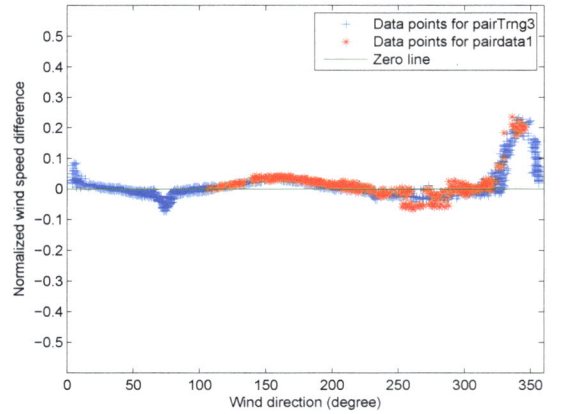

Figure 9. Significant overlap of patterns extracted from pair-data1 and pairTrng3.

as follows:

$$\rho(q,p) = \frac{\langle \overline{S}_q, \overline{S}_p \rangle}{\|\overline{S}_q\|_2 \|\overline{S}_p\|_2} \quad (5)$$

Here, $\langle \cdot, \cdot \rangle$ stands for the inner product of two vectors. The larger the correlation coefficient, the more similar the two shapes of p and q are. The training profile p^* is selected for comparison with test file q if

$$p^* = \arg\min_p \left(\frac{Dis(q,p)}{\|\overline{S}_q\|_2} + \sqrt{1 - \rho^2(q,p)} \right) \quad (6)$$

The objective function is the average of the distance measure and shape measure.

2.5. Decision Making

There are four possible conditions of the paired anemometers in the test data: both are normal (0), anemometer 1 fails (1), anemometer 2 fails (2), and both fail (3). Following assumptions are made regarding these four conditions:

(0) If both anemometers work normally, the feature, i.e., the relation between the wind speed difference and the wind direction, should be very similar to its corresponding matched training pattern. That is, the feature extracted from the test data file will have a significant overlap with the corresponding training data pattern. Figures 9[1] is an example.

(1) It is assumed that if an anemometer fails, its reading is generally smaller than the true value, especially for mechanical failures. Based on the definition of the wind speed difference in Eq. (1), if anemometer 1 fails, the pattern will have a downward shift. Namely, the wind speed difference values will take more negative values with the change of wind direction.

(2) With the same assumption, if anemometer 2 fails, the pattern shows an upper shift. That is, the wind speed difference values will take more positive values. There are many such kinds of patterns in test data, which shows that this assumption may be right. Figures 10 and 11 are the examples of these two conditions.

(3) If both anemometers fail, the pattern is not predictable, i.e., it does not show any of the above characteristics in an obvious way.

To make a decision for each test file, the following algorithm is designed taking into account the above assumptions. Assume that for test file q, training file p is selected for comparison through the pattern search step. For wind direction d where wind speed difference values are available in both training and test data, define $S_p(d)\_min = \min\{S_{p_1}(d), \cdots, S_{p_n}(d)\}$ and $S_p(d)\_max = \max\{S_{p_1}(d), \cdots, S_{p_n}(d)\}$, assuming that there are n wind speed difference values at wind direction d in training file p. Then for test data q and for the same direction d, count the number of data points in, above or below the range $[S_p(d)\_min, S_p(d)\_max]$, which are denoted as $C_{q,in}(d)$, $C_{q,above}(d)$, and $C_{q,below}(d)$, respectively. There are two ways to proceed based on these counts. One is to make a decision for each wind direction and fuse these decisions to generate a global decision (decision fusion). The other one is to add up the total number of data points in, above or below the normal ranges and make a decision based on that (data fusion). Since we have no ground truth and the characteristics of wind speed difference vary for different wind directions, we develop the following hybrid method. The whole 360° is divided into 36 bins. The counts in each bin add up, i.e.,

$$C_{q,xx}(i) = \sum_{d \in Bin_i} C_{q,xx}(d) \quad (7)$$

---

[1]Figures 9, 10, and 11 can be viewed better with a color print.

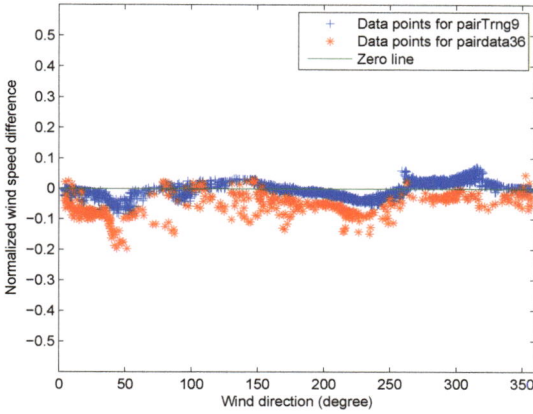

Figure 10. Significant downward shift of the pattern extracted from pairdata36 compared to the pattern from pairTrng9.

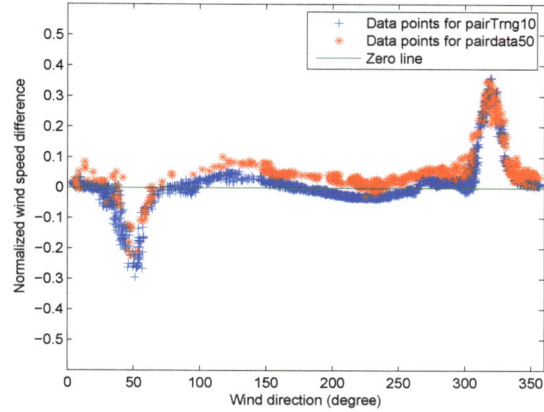

Figure 11. Upper shift of the pattern extracted from pairdata50 compared to the pattern from pairTrng10.

where $i = 1, \cdots, 36$ and xx can be *in*, *above* and *below*. Decision is made for each bin using the following rule:

$$U_q(i) = \begin{cases} 0 & \text{if } C_{q,in}(i) > T_i \\ 1 & \text{if } C_{q,below}(i) > T_i \\ 2 & \text{if } C_{q,above}(i) > T_i \\ 3 & \text{otherwise} \end{cases} \quad (8)$$

where threshold $T_i = \frac{C_{q,in}(i) + C_{q,above}(i) + C_{q,below}(i)}{2}$. That is, whichever of the first three conditions dominating indicates the condition of that bin. If there is no one that dominates, decision 3 is made. The majority of local decisions is chosen as the global decision. This hybrid method can not only smooth out the noise effect, but also preserve the variation of data pattern in different directions. Note that if no data points exist in some bins, those bins do not participate in decision making.

2.6. Results and Discussion

In the competition, the results are evaluated based on whether the proposed algorithm can accurately determines the conditions of the paired anemometers for each test file. Credit for each file is gained only if the decisions for both anemometers are correct. Visualization of our results for paired data is provided in Figure 12 on the top of next page. Condition indicators 0, 1, 2, and 3 are defined in Section 2.5. There are a total of 287 test files with decision 0, 43 files with decision 1, 39 files with decision 2, and 51 files with decision 3.

For paired data analysis, the normalized wind speed difference (NWSD) as a function of the wind direction is extracted as a main feature for the purpose of faulty anemometer detection. Since wind data are collected from different environments, under different weather conditions and with different tower configurations, taking the difference and normalization of paired data can reduce environmental impacts effectively,

while the NWSD pattern with respect to the wind direction can help identify similar anemometer configurations, thus putting training/testing-file comparison and anomaly detection in the same context. If raw data instead of the proposed feature is used, almost all test files look different/erroneous compared to the training files.

3. METHODOLOGY FOR SHEAR DATA ANALYSIS

For shear data, the problem is to decide whether all of an array of anemometers work normally. Similarly, a data preprocessing step has to be taken to eliminate some obviously useless data. Specifically, the measurement range test is conducted. It should be noted that the effect of icing conditions in cold climate is huge so that a majority of data are under the influence to different extents (Schaffner, 2002). For this problem, the same criteria as specified for paired data are used to partially mitigate the icing effect.

3.1. Irregular Data Elimination

Generally, the wind speed increases with the height because of the wind shear effect. However, in the training data with all anemometers in a normal condition, there exist many measurements violating this rule. This indicates that the measurements do not always reflect the true wind speeds, which may be due to the environmental factors rather than anemometer failures. We define a record containing this kind of measurements as irregular data and they make the detection problem more challenging. In Table 1, we summarize the mean temperature and the percentage of irregular data for all 7 shear training files. It is noted that the ones with lower temperatures generally have more irregular data. Thus, the irregularity is more likely the result of icing effects. To reduce the effect of icing on decision making, we eliminate all the irregular data.

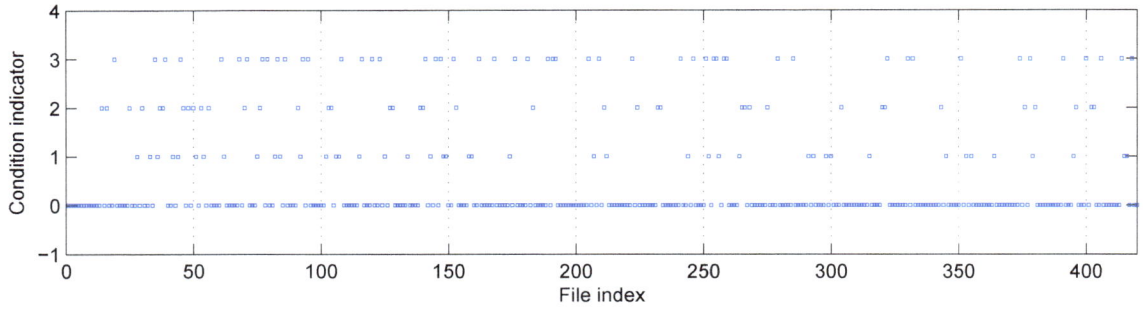

Figure 12. Results for paired data.

ShearTrng file	Mean Temperature (°F)	Irregular Data (%)
1	11.45	11
2	50.81	8
3	45.20	3
4	3.07	67
5	3.23	68
6	11.02	35
7	10.93	35

Table 1. Mean temperature and the percentage of irregular data for shear training files.

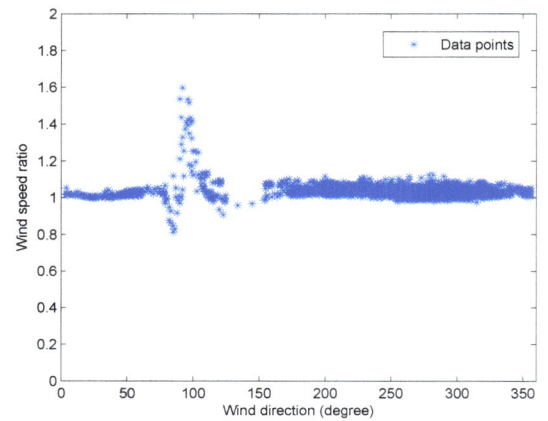

Figure 13. Wind speed ratio between 59m and 51m versus wind direction for shearTrng1.

As mentioned in Section 2, the configuration of the tower also has effects on wind speed measurements. In Figure 13, the ratio of wind speeds at 59m and 51m as a function of the wind direction is plotted. It shows that around 90°, the direction in which the anemometers are installed, the ratio takes significantly different values. The measurements around the wind directions, to which the anemometers are pointed, fail to reflect the normal situation and therefore are eliminated.

3.2. Model fitting

After eliminating irregular data due to icing and/or the tower, we assume the failure of anemometers is the dominating factor of irregular patterns in test data, if any. One widely used wind shear model is a power law model (Burton, Sharpe, Jenkins, & Bossanyi, 2001), and is given as follows,

$$\frac{s}{s_r} = \left(\frac{h}{h_r}\right)^\alpha \qquad (9)$$

where s is the wind speed at some specific height h, s_r the wind speed at a reference height h_r, and α the shear exponent. If we use the shear data to fit the model, we expect that the sum of squared residuals (SSR) tends to be small for normal data while be relatively large for abnormal data. Since the wind speed changes across time and space, to make the SSR

comparable for different files, normalization by the maximum value of an array of wind speeds at each time is taken. As a result, all normalized wind speed fall into the range of $[0, 1]$. Figure 14 shows an example of the fitted power law model and normalized sample data points for shearTrng1. SSR is used as a performance measure of the given shear data.

3.3. Decision Making

There are three types of shear data files: three anemometers at (57m, 45m, 35m), four anemometers at (59m, 51m, 30m, 10m) and four anemometers at (49m, 39m, 30m, 10m). For the first two types, the training and test data have very similar temperature. Since there is only one and two training files for these two types of data respectively, a 25-day training data file is divided into 5 files with 5 days of data each. The SSRs are calculated as shown in Figure 15 for the four-anemometer configuration for 5 smaller training files and 20 test files. The decision making rule is as follows. The average of five SSR values from training data is used as a threshold, partially eliminating the randomness such as noise. If

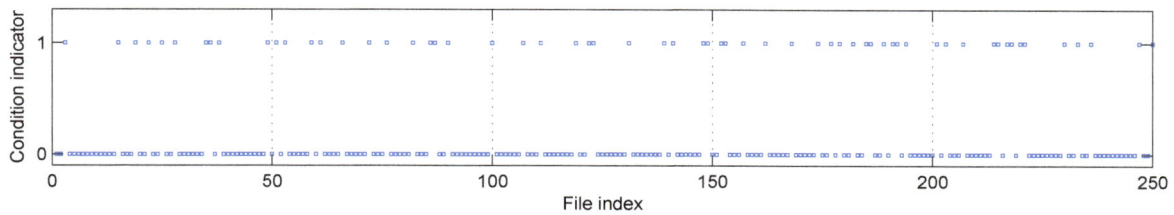

Figure 16. Results for shear data

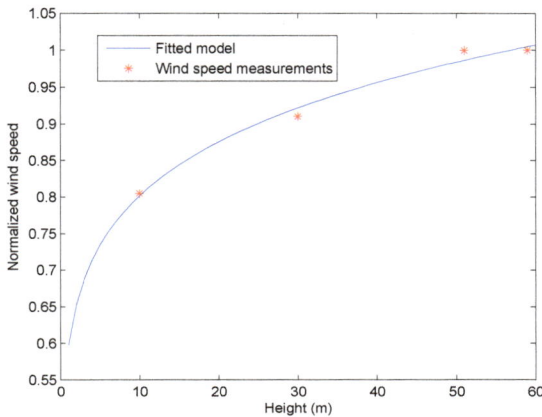

Figure 14. Normalized wind speeds measurements and fitted wind speeds using a power law model.

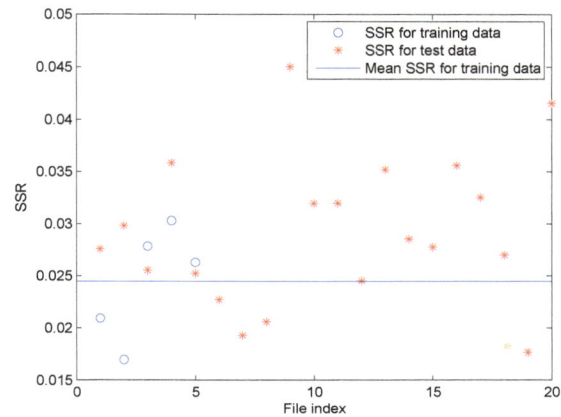

Figure 15. SSR for training and testing data for the four-anemometer configuration (59m, 51m, 30m, 10m).

the SSR of a test file is greater than this threshold, the decision that not all anemometer working normally is made. For the configuration with (49m, 39m, 30m, 10m), there are two types of temperature values: below the icing points and much above the icing points. This fact motivates us to compare the test data with the training files of similar temperature. The rest of algorithm remains the same.

3.4. Results and Discussion

In the competition, the results are evaluated based on whether the proposed algorithm can accurately determine the condition of an array of anemometers for each test file. Credit for each file is gained if the decision about whether any faulty anemometer occurs among the array of anemometers is correct. Visualization of our results for shear data is provided in Figure 16. A test file without any faulty anemometer is indicated with number 0, otherwise with number 1. There are a total of 193 files with decision 0 and 62 files with decision 1.

For shear data, the performance of the proposed algorithm largely depends on the elimination of noisy and irregular data. The sum of squared residuals (SSR) after fitting a power law model for an array of normalized wind speed measurements is used as the main feature to detect if any fault exists in the

array. The assumption is that a faulty array tends to have a larger SSR compared to that of training data of the same configuration. Data of different time of a day used for model fitting may influence the decision. This is because the wind shear is also a function of the time of a day and exhibits the diurnal variation, i.e., the wind shear exponent in the daytime is significantly smaller than at night (K. Smith, Randall, Malcolm, Kelley, & Smith, 2002). Therefore, an improved feature may be SSR as a function of the time of a day, the pattern variation of which can also be used as an indicator of possible faults. This will be investigated in our future work.

4. CONCLUSIONS

In this paper, we have developed a series of methods including data preprocessing, feature extraction and pattern identification to solve the anemometer condition diagnosis problem of the PHM 2011 Data Challenge Competition. The main idea of the algorithms is to extract useful features showing diecernable patterns of training and test data so that they can reflect the health condition of anemometers. Since the data patterns may also be significantly influenced by various factors such as icing and the tower rather than anemometer failures, considerable efforts have been taken for elimination of irregular data due to these environmental factors. For paired data, the relation between the normalized wind speed difference and the wind direction is used as the key feature for pattern identification and decision making. For shear data, the sum of squared residuals after model fitting is used for decision making. Several important assumptions are made for algorithm development, some of which have been justified by our observation of the data and the domain knowledge.

There are several aspects that we can pursue to further improve the diagnosis performance: the development of more efficient methods to reduce environmental effects and eliminate outliers, e.g., a new criterion by looking at wind direction to determine the range of useless data; extraction of features that are more sensitive to anemometer failures. The influence from environment is a major challenge which prevents us from accurately capturing the characteristics of anemometer failures. On the other hand, features more sensitive to anemometer failures would lead to a higher failure detection probability and a lower false alarm rate. These efforts will all have great practical values to the wind energy development.

ACKNOWLEDGMENT

This work was supported by NSF under Grant CPS-0932297.

REFERENCES

Barnett, V., & Lewis, T. (1994). *Outliers in Statistical Data (3rd Edition)*. John Wiley and Sons.

Basu, S., & Meckesheimer, M. (2007). Automatic Outlier Detection for Time Series: An Application to Sensor Data. *Knowledge and Information Systems, 11*(2), 137–154.

Beltran, J., Llombart, A., & Guerrero, J. J. (2009a). A Bin Method with Data Range Selection for Detection of Nacelle Anemometers faults. In *Proceedings of European Wind Energy Conference and Exhibition (EWEC)*. March 16-19, Marseille, France,.

Beltran, J., Llombart, A., & Guerrero, J. J. (2009b). Detection of Nacelle Anemometers Faults in a Wind Farm. In *Proceedings of International Conference on Renewable Energies and Power Quality (ICREPQ)*. April 15-17, Valencia, Spain.

Burton, T., Sharpe, D., Jenkins, N., & Bossanyi, E. (2001). *Wind Energy Handbook (2nd Edition)*. Wiley.

Chan, P., & Mahoney, M. (2005). Modeling Multiple Time Series for Anomaly Detection. In *Proceedings of Fifth IEEE International Conference on Data Mining* (pp. 90–97). November 27-30, Houston, TX, USA.

Chandola, V., Banerjee, A., & Kumar, V. (2009). Anomaly Detection: a Survey. *ACM Computing Surveys, 41*(3), 1–58.

Delfino, T. N., Puttini, L. C., & Galvao, R. K. H. (2010). Fault Prognosis of an Air Flow Sensor. In *Proceedings of XVIII Congresso Brasileiro de Automtica (CBA)*. September 12-16. Bonito, MS, Brazil.

Duda, R. O., Hart, P. E., & Stork, D. G. (2000). *Pattern Classification (2nd Edition)*. Wiley-Interscience.

IEA. (1999). *Annex XI: Recommended Practices for Wind Turbine Testing and Evaluation 11. Wind Speed Measurement and Use of Cup Anemometry, 1.* Paris: IEA.

Kenyon, P. R., & Blittersdorf, D. C. (1996). *Accurate Wind Measurements in Icing Environments, Solutions to the Problem of Invalid Data from Frozen Anemometers and Direction Vanes.* Report NRG System.

Knorr, E. M., Ng, R. T., & Tucakov, V. (2000). Distance-based Outliers: Algorithms and Applications. *The Very Large Data Bases (VLDB) Journal, 8*(3-4), 237–253.

Kusiak, A., Zheng, H., & Zhang, Z. (2011). Virtual Wind Speed Sensor for Wind Turbines. *Journal of Energy Engineering, 137*(2), 59-69.

Lubitz, W. D. (2009). Effects of Tower Shadowing on Anemometer Data. In *Proceedings of 11th Americas Conference on Wind Engineering*. June 22-26, San Juan, Puerto Rico.

Schaffner, B. (2002). *Wind Energy Site Assessment in Harsh Climatic Conditions, Long Term Experience in Swiss Alps*. Report METEOTEST.

Siegel, D., & Lee, J. (2011). An Auto-Associative Residual Processing and K-means Clustering Approach for Anemometer Health Assessment. *International Journal of Prognostics and Health Management, 2*(2).

Smith, K., Randall, G., Malcolm, D., Kelley, N., & Smith, B. (2002). Evaluation of Wind Shear Patterns at Midwest Wind Energy Facilities. In *Proceedings of the American Wind Energy Association (AWEA) Windpower 2002 Conference*. June, Portland, OR, USA,.

Smith, R., Bivens, A., Embrechts, M., Palagiri, C., & Szymanski, B. (2002). Clustering Approaches for Anomaly Based Intrusion Detection. In *Proceedings of Intelligent Engineering Systems through Artificial Neural Networks*. (pp. 579–584). ASME Press.

Longji Sun received his B.E. degree in communication engineering from University of Shanghai for Science and Technology, Shanghai, China in 2010. He is now pursuing his Ph.D. degree in the School of Electrical and Computer Engineering at Oklahoma State University. His current research interests include structural health monitoring using wireless sensor networks, cooperative optimization and cognitive radios. He is a student member of IEEE.

Chao Chen received his B.E. degree in automation control systems from Wuhan University of Technology, Wuhan, China in 2007. From 2007 to 2008, he worked as an electrical system design engineer at Shanghai Waigaoqiao Shipbuilding Company, Ltd., Shanghai, China. He is currently an M.S student in the School of Electrical and Computer Engineering at Oklahoma State University. His research interests include structural health monitoring, signal processing, information fusion and statistical analysis.

Qi Cheng received the B.E. degree in electrical engineering (highest honors) from Shanghai Jiao Tong University, Shanghai, China, in July 1999, and the M.S. and Ph.D. degrees in electrical engineering from Syracuse University, Syracuse, NY, in 2003 and 2006, respectively. From 1999 to 2000, she worked as a System Engineer at Guoxin Lucent Technologies Network Technologies Company, Ltd., Shanghai, China. Since August 2006, she has been with Oklahoma State University, as an Assistant Professor in the School of Electrical and Computer Engineering. Her area of interest mainly focuses on statistical signal processing and data fusion with applications in wireless communications and distributed sensor networks. She is a member of IEEE.

Condition Monitoring Method for Automatic Transmission Clutches

Agusmian Partogi Ompusunggu[1], Jean-Michel Papy[2], Steve Vandenplas[3], Paul Sas[4], and Hendrik Van Brussel[5]

[1,2,3] *Flanders' Mechatronics Technology Centre (FMTC), Celestijnenlaan 300D, 3001 Heverlee, Belgium.*
agusmian.ompusunggu@fmtc.be
jean-michel.papy@fmtc.be
steve.vandenplas@fmtc.be

[4,5] *K.U.Leuven, Department of Mechanical Engineering, Division PMA, Celestijnenlaan 300B, 3001 Heverlee, Belgium.*
paul.sas@mech.kuleuven.be
hendrik.vanbrussel@mech.kuleuven.be

ABSTRACT

This paper presents the development of a condition monitoring method for wet friction clutches which might be useful for automatic transmission applications. The method is developed based on quantifying the change of the relative rotational velocity signal measured between the input and output shaft of a clutch. Prior to quantifying the change, the raw velocity signal is preprocessed to capture the relative velocity signal of interest. Three dimensionless parameters, namely the normalized *engagement duration*, the normalized *Euclidean distance* and the *spectral angle mapper distance*, that can be easily extracted from the signal of interest are proposed in this paper to quantify the change. In order to experimentally evaluate and verify the potential of the proposed method, clutches' life data obtained by conducting accelerated life tests on some commercial clutches with different lining friction materials using a fully instrumented SAE#2 test setup, are utilized for this purpose. The aforementioned parameters extracted from the experimental data clearly exhibit progressive changes during the clutch service life and are well correlated with the evolution of the *mean* coefficient of friction (COF), which can be seen as a reference feature. Hence, the quantities proposed in this paper can therefore be seen as principle features that may enable us to monitor and assess the condition of wet friction clutches.

1. INTRODUCTION

Vehicles equipped with automatic transmissions have gained popularity in recent years. As is obvious from its name, an automatic transmission is a transmission which shifts power or speed by itself. In this kind of transmissions, wet friction clutches are one of critical components that play a major role on the performance.

Wet friction clutches are machine elements enabling the power transmission from an input shaft (driving side) to an output shaft (driven side) during the operation, based on the friction occurring on lubricated contacting surfaces. The contacting surfaces comprise friction surface (friction disc) and counter surface (separator disc). The clutch is lubricated by an automatic transmission fluid (ATF) having a main function as a cooling lubricant cleaning the contacting surfaces and giving smoother performance and longer life. Besides for the clutch lubrication, this oil is also used for the clutch actuation.

The presence of the ATF in the clutch, however, reduces the coefficient of friction (COF). In applications where high power is mandatory, *e.g.* heavy duty vehicles (tractors, harvesters, *etc*), the clutch is therefore designed with multiple friction and separator discs. This configuration is known as a multi-disc wet friction clutch as schematically shown in Figure 1, in which the friction discs are mounted to the hub by splines, and the separator discs are mounted to the drum by lugs. The friction disc is made of a steel-core disc with lining friction material bonded on both sides and the separator disc is made of plain steel. In addition, the input shaft is commonly connected to the drum side, while the output shaft is connected to the hub side.

An electro-mechanical-hydraulic actuator is commonly used for both disengagement and engagement mechanisms of wet friction clutches. This actuator consists of some main components, such as: a piston, a returning spring which is always under compression and a hydraulic group consisting of a control valve, an oil pump, a filter, *etc*. Figure 1 shows the assembly of the piston and the returning spring in the interior of a wet friction clutch. To engage a wet friction clutch, a pressur-

(a)

(b)

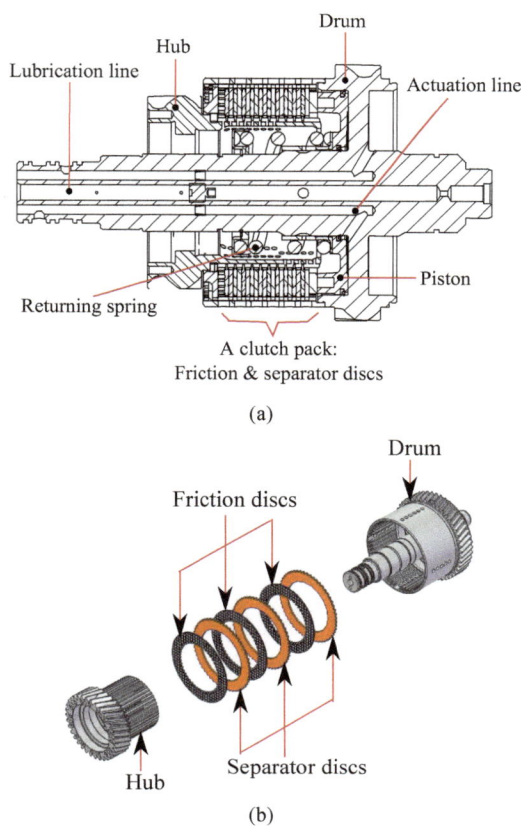

Figure 1. Configuration of a multi-disc wet friction clutch, (a) cross-sectional and (b) exploded view.

ized ATF actuated by the control valve is applied through the *actuation line* in order to generate a force acting on the piston. When the applied pressure exceeds a certain value to overcome the resisting force arising from both spring force and frictional force occurring between the piston and the internal surface of the drum, the piston starts moving and eventually pushes both friction and separator discs toward each other. To disengage the clutch, the pressurized ATF is released such that the returning spring is allowed to push the piston back to its rest position.

An unexpected failure occurring in the clutch can therefore lead to total breakdown of the vehicles. The impact can put human safety at risk, possibly cause long term vehicle down times, and result in high maintenance costs. In order to minimize the negative impacts caused by an unexpected breakdown, an optimal maintenance strategy driven by an accurate condition monitoring and prognostics needs to be applied for wet friction clutches. Although they are critical components, to our knowledge very little attention has been paid to these particular components in terms of the development of a condition monitoring method (tool). The main objective of this study is to develop a condition monitoring method for wet

friction clutches that can be practically used in real-life applications. Furthermore, the developed method can be possibly extended towards clutch condition prognosis, but this is out of the scope of the paper.

Feature extraction is a key step to succeed in the development of a condition monitoring method. A feature can be derived based on the physics of degradation of the case of interest or based on heuristic (data-driven) approach. The derivation of physics-based features requires a profound understanding about the physics of degradation, while the derivation of heuristic-based features requires a large number of training data and experience about the case of interest. Furthermore, a parameter or quantity can be considered as a principle feature if it effectively delivers useful information about the failure mode and level. In general, the evolution of (a combination of) principle features can be associated with the progress of a target failure.

The coefficient of friction (COF), which can be seen as a physics-based feature, has been used for many years as a principle feature for monitoring the condition of wet friction clutches (Matsuo & Saeki, 1997; Ost, Baets, & Degrieck, 2001; Maeda & Murakami, 2003; Li et al., 2003; Fei, Li, Qi, Fu, & Li, 2008). However, the use of the COF for clutch monitoring is possibly expensive and not easily implementable for real-life applications, due to the fact that at least two sensors are required to extract the COF, namely (i) a *torque sensor* and (ii) a *force sensor*, which are commonly difficult to install in a transmission (*i.e.* typically not available in today's transmissions). The quasi-steady-state clutch torque may be estimated from the torque-velocity characteristics of engine/torque converter, while the normal (axial) clutch force may be approximated from the pressure applied on the piston. However, the torque and normal force estimations with this approach can lead to inaccurate COF estimation.

Furthermore, several methods have been proposed in literature for assessing the condition of wet friction clutches based on the quality of the friction material, namely (i) Scanning Electron Microscope (SEM) micrograph, (ii) surface topography, (iii) Pressure Differential Scanning Calorimetry (PDSC) and (iv) Attenuated Total Reflectance Infrared spectroscopy (ATR-IR) (Jullien, Meurisse, & Berthier, 1996; Guan, Willermet, Carter, & Melotik., 1998; Li et al., 2003; Maeda & Murakami, 2003; Nyman, Maki, Olsson, & Ganemi, 2006). Nevertheless, these methods are not practically implementable during operation, owing to the fact that the friction discs have to be taken out from the clutch pack and then prepared for assessing the degradation level. In other words, an online condition monitoring system can not be realized by using these existing methods.

In our recent studies (Ompusunggu, Papy, Vandenplas, Sas, & VanBrussel, 2009; Ompusunggu, Sas, VanBrussel, Al-Bender, Papy, & Vandenplas, 2010; Ompusunggu, Sas, Van-

Brussel, Al-Bender, & Vandenplas, 2010; Ompusunggu, Sas, et al., 2011; Ompusunggu, Papy, Vandenplas, Sas, & Van-Brussel, n.d.), some potential features extracted from the pre- and post-lockup torsional vibration and normal-mode vibration signals, have been investigated and proven to be relevant for condition monitoring of wet friction clutches. These potential features are derived based on both physical and heuristic reasonings. Although they have potential as features for clutch monitoring, determining principle features that are robust, relatively easy to measure and inexpensive to compute for clutch condition monitoring remains a subject for further investigation.

As the degradation occurring in wet friction clutches progresses, the frictional characteristics change which consequently alters the behavior and performance of clutches. Based on this reasoning, it is hypothesized that the change of clutch behavior is also reflected by the change of the relative rotational velocity signal measured between the input and output shaft of the clutch. This hypothesis has been confirmed theoretically and experimentally in another recent study (Ompusunggu, Janssens, Al-Bender, Sas, & VanBrussel, 2011). In this present paper, the development of a condition monitoring method for wet friction clutches based on monitoring the change of the relative velocity signal is presented. Three parameters, namely the engagement duration and two dissimilarity measures, namely the Euclidean distance and Spectral Angle Mapper (SAM) distance (Kruse et al., 1993; Paclik & Duin, 2003) are proposed in this paper as features for clutch monitoring. In order to experimentally evaluate and verify the potential of the proposed method, clutches' life data obtained by conducting accelerated life tests on some commercial clutches with different lining friction materials using a fully instrumented SAE#2 test setup, are utilized for this purpose. In the tests, the COF is measured and used as a *reference feature* to evaluate the relevance of the proposed features.

The remainder of this paper is organized as follows. After introducing the objective and motivation, the methodology of clutch monitoring developed in this study is presented and discussed in Section 2. Service life data of some clutches obtained from accelerated life tests carried out on the used SAE#2 test setup are employed for the evaluation of the method, where the experimental aspects are described in Section 3. The results obtained after applying the proposed method to the clutches' life data are further presented and discussed in Section 4. Finally, some conclusions drawn from the study are presented in Section 5.

2. METHODOLOGY

Background that motivates the development of the methodology is first discussed in this section. Then, the signal preprocessing technique prior to computing the proposed features is described. Finally, the formulas to compute the features are presented and discussed.

2.1. Background

As reported in literature, degradation occurring in wet friction clutches (*e.g.* friction materials) alters the frictional characteristics. A change of the frictional characteristics during the clutches' service life is exposed by a decreasing coefficient of friction (COF) (Ost et al., 2001; Fei et al., 2008). With the same operational condition, this decreasing COF implies that the torque transmitted during clutch engagement drops. Besides a COF reduction, another aspect which is often associated with the change of the clutch frictional characteristics is the loss of anti-shudder property. Due to this, the damping characteristic of the clutch system changes (the equivalent damping can be negative). It is commonly accepted that the loss of anti-shudder property in a clutch can lead to the occurrence of the stick-slip or/and self excited vibration in the driveline.

As a result of the change of the frictional characteristics, the dynamic responses of clutches during and after the engagement phase also change. In this study, the relative velocity profile measured between the input and output shafts of a clutch, which can be seen as a representation of the dynamic behavior during the engagement phase, is considered as the one that is significantly affected by the change of the clutch frictional characteristics. It is important to note that the use of the latter signal is motivated by the fact that it is possible to measure (online) in real-life applications since the rotational velocity sensors (*e.g.* Hall-effect encoders) are typically available in automatic transmissions.

2.2. Capturing the signal of interest

Prior to computing the features, the raw signals obtained from measurements first need to be preprocessed. Figure 2 graphically illustrates the signal preprocessing step, namely the procedure to capture the *relative rotational velocity signal of interest* based on two raw signals: (i) the *relative rotational velocity* signal and (ii) *pressure applied to the clutch piston* signal. Once the signal of interest is captured using the two raw signals, features can then be extracted for clutch monitoring purpose.

Let the signal of interest be captured at a given duty cycle with a predetermined time record length τ, and suppose that the time record length is kept the same for all duty cycles. For the sake of consistency, the signal is always captured at the *same* reference time instant. It is reasonable to consider the time instant when the ATF pressure applied to the clutch pack $p(t)$ starts to increase from zero value t_f as the reference time instant. For an ideal pressure signal, *i.e.* continuous and noiseless, t_f can be mathematically formulated as:

$$t_f = \min \{\forall t \in \mathbb{R} : \quad p(t) > 0\}. \tag{1}$$

While the applied pressure is increasing, contact is gradually established between the separator and friction discs. As a result, the transmitted torque increases, while the relative velocity $n_{rel}(t)$ decreases. The clutch is fully engaged when the relative velocity reaches zero value for the first time at the lockup time instant. For an ideal relative velocity signal, the lockup time instant t_l can be formulated in a similar way as in Equation (1):

$$t_l = \min \{\forall t \in \mathbb{R} : \quad n_{rel}(t) = 0\}. \qquad (2)$$

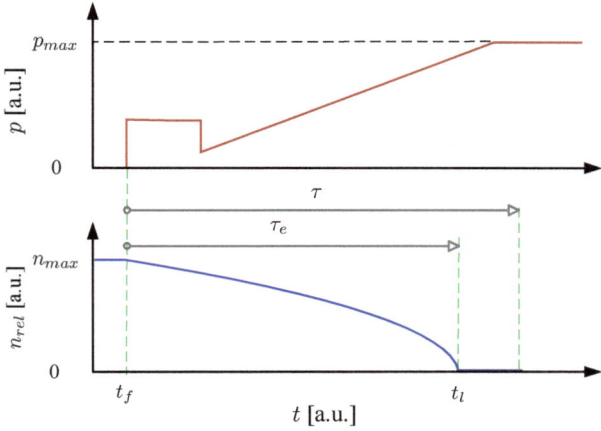

Figure 3. A graphical representation illustrating the estimation of (a) the reference time instant $\hat{t}_f$ based on the *pressure signal* p and (b) the lockup time instant $\hat{t}_l$ based on the *relative velocity signal* n_{rel}.

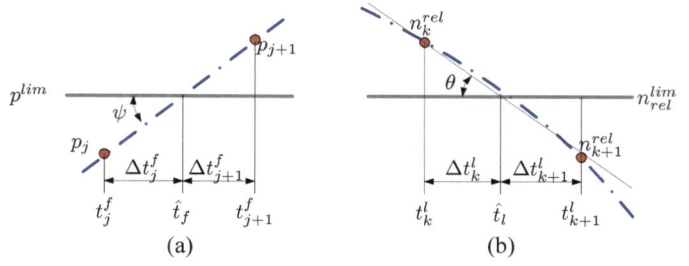

Figure 2. A graphical illustration of how to capture the relative velocity signal of interest. The upper and lower figures respectively denote the typical pressure and the raw relative velocity signal. Note that a.u. is the abbreviation of arbitrary unit.

In practice, data obtained from measurements are normally discretized with some possible noise. As a result, the estimation of t_f and t_l is no longer straightforward. To deal with this, two simple algorithms based on the linear interpolation technique are developed. The interpolation algorithms identify both time instants t_f and t_l, when limit values for the pressure p^{lim} and the velocity n_{rel}^{lim} are exceeded, as illustrated in Figure 3. The limit values are chosen such that they are higher than the floor noise level of pressure and velocity, i.e. $p^{lim} > \max \{n_p^f\}$ and $n_{rel}^{lim} > \max \{n_n^f\}$, with n_p^f and n_n^f respectively denoting the pressure and the velocity floor noise. The algorithms are discussed in the subsequent paragraphs.

Assume that each measured signal is discretized with sampling time T_s and number of sampling points N. By knowing that a discrete signal is a set of finite sampling points, it is therefore convenient to treat a measured signal as a vector. Let two vectors $\boldsymbol{P}$ and $\boldsymbol{V}$ be respectively denoting the discrete raw pressure and velocity signal vectors with N sample points. Hence, these N-dimensional vectors can be expressed

in a matrix format as follows:

$$\boldsymbol{P} = [p_1, p_2, \ldots p_j, \ldots, p_{N-1}, p_N]^T, \qquad (3)$$
$$\boldsymbol{V} = [n_1^{rel}, n_2^{rel}, \ldots, n_k^{rel}, \ldots, n_{N-1}^{rel}, n_N^{rel}]^T. \qquad (4)$$

Equations (1) and (2) do not apply anymore for discrete and noisy (actual) signals. Nevertheless, these two equations can be adapted in order to estimate the reference time instant $\hat{t}_f$ and lockup time instant $\hat{t}_l$ from actual signals. The index j of the instantaneous applied pressure p_j just before crossing the p^{lim}, see Figure 3(a), is computed with the following equation:

$$j = \min \{\forall j \in \mathbb{Z} : \quad (p_j - p^{lim}) * (p_{j+1} - p^{lim}) < 0\}. \qquad (5)$$

Based on the index j, the estimated reference time instant $\hat{t}_f$ can be computed as follows, see Figure 3(a):

$$\hat{t}_f = t_j^f + \Delta t_j^f \quad \text{or} \quad \hat{t}_f = t_{j+1}^f - \Delta t_{j+1}^f \qquad (6)$$

with:

$$t_j^f = (j-1)T_s \quad \text{and} \quad t_{j+1}^f = jT_s,$$
$$\Delta t_j^f = \left(\frac{p^{lim} - p_j}{p_{j+1} - p_j}\right) T_s,$$
$$\Delta t_j^f = \left(\frac{p_{j+1} - p^{lim}}{p_{j+1} - p_j}\right) T_s.$$

The index k of the relative velocity signal n_{rel}^k just before crossing the n_{rel}^{lim} is computed with the following equation:

$$k = \min \{\forall k \in \mathbb{Z} : \quad (n_k^{rel} - n_{rel}^{lim}) * (n_{k+1}^{rel} - n_{rel}^{lim}) < 0\}. \qquad (7)$$

In similar way, based on the index k, the estimated lockup time instant $\hat{t}_l$ can be computed as follows:

$$\hat{t}_l = t_k^l + \Delta t_k^l \quad \text{or} \quad \hat{t}_l = t_{k+1}^l - \Delta t_{k+1}^l \qquad (8)$$

with:

$$t_k^l = (k-1)T_s \quad \text{and} \quad t_{k+1}^l = kT_s,$$

$$\Delta t_k^l = \left(\frac{n_{rel}^k - n_{rel}^{lim}}{n_{rel}^k - n_{rel}^{k+1}} \right) T_s,$$

$$\Delta t_k^l = \left(\frac{n_{rel}^{lim} - n_{rel}^{k+1}}{n_{rel}^k - n_{rel}^{k+1}} \right) T_s.$$

2.3. Feature extraction

Formal definitions of the developed features (engagement duration, Euclidean distance and Spectral Angle Mapper distance) and the mathematical expressions to compute them are discussed in this subsection. The first two features are dimensional quantities while the third one is dimensionless. The first two features are normalized such that they become dimensionless quantities and are in the same order of magnitude as of the third feature.

2.3.1. Engagement Duration

By definition, the engagement duration τ_e is referred to as the time interval between the lockup time instant t_l and the reference time instant t_f, as graphically illustrated in Figure 2. However, these two *ideal* time instants cannot be obtained from the actual measurement data. Instead, these time instants are estimated based on the procedure previously described. Once both estimates time instants $\hat{t}_f$ and $\hat{t}_l$ have been determined, the estimate engagement duration $\hat{\tau}_e$ can then be simply computed as follows:

$$\hat{\tau}_e = \hat{t}_l - \hat{t}_f. \tag{9}$$

Without loss of generality, $\hat{\tau}_e$ can be normalized with respect to the engagement duration measured at the first cycle (initial condition) $\hat{\tau}_e^r$, according to the following equation:

$$\bar{\tau}_e = \frac{\hat{\tau}_e - \hat{\tau}_e^r}{\hat{\tau}_e^r}, \tag{10}$$

where $\bar{\tau}_e$ denotes the dimensionless engagement duration.

2.3.2. Dissimilarity Measures

A dissimilarity measure is a metric that quantifies the dissimilarity between objects. For the sake of condition monitoring, the dissimilarity measure between an object that represents an arbitrary condition and the reference object that represents a healthy condition, can be treated as a feature. Thus, the dissimilarity measure between two identical objects is (close to) zero; the dissimilarity measure between two non-identical objects on the other hand is not zero. Here, the object refers to the relative velocity signal. Two dissimilarity measures, namely the Euclidean distance and the Spectral Angle Mapper (SAM) distance, are considered in this paper because of their computational simplicity.

The basic principle behind the dissimilarity approach is that the measured signals of interest are treated as vectors. Let X be a K dimensional vector, $x_i, i = 1, 2, \ldots, K$, denoting the discrete signal of the relative velocity measured in an initial (healthy) condition and Y be a K dimensional vector, $y_i, i = 1, 2, \ldots, K$, denoting the discrete signal of the relative velocity measured in an arbitrary condition. The vector X representing a healthy condition is referred to as the "baseline". Note that one can also take the average of several relative velocities measured in the healthy condition as the baseline, in order to increase the confidence level and accuracy.

The Euclidean distance (D_E) between the vectors X and Y is defined as:

$$D_E(X, Y) = \sqrt{\sum_{i=1}^{K} (x_i - y_i)^2}. \tag{11}$$

For convenience, the Euclidean distance D_E can be normalized in accordance with the following equation:

$$\bar{D}_E(X, Y) = \frac{D_E(X, Y)}{x_1 \sqrt{K}}, \tag{12}$$

where $\bar{D}_E$ denotes the dimensionless Euclidean distance and $x_1 = \max\{X\} > 0$ denotes the maximum value of the *baseline*, *i.e.* the initial relative velocity in healthy condition. This way, the dimensionless Euclidean distance $\bar{D}_E$ is bounded between 0 and 1, see Appendix A.

The SAM distance is a measure of the angle between two vectors and is therefore dimensionless. Mathematically, the SAM distance $\bar{D}_{SAM}$ between the vectors X and Y is defined as:

$$\bar{D}_{SAM}(X, Y) = \cos^{-1} \left(\frac{\sum_{i=1}^{K} x_i y_i}{\sqrt{\sum_{i=1}^{K} x_i^2} \sqrt{\sum_{i=1}^{K} y_i^2}} \right). \tag{13}$$

Recall that the distance from an object to itself is zero and that a distance is always non-negative. To compute the two dissimilarity measures, a baseline X, *i.e.* the signal of interest in healthy condition, is required and the signal of interest in an arbitrary condition Y *must* have the same size with the baseline. In this paper, the signal of interest from the first duty cycle is taken as the baseline since it can represent the healthy condition.

3. EXPERIMENTS

Service life data of wet friction clutches are required for the evaluation of the developed condition monitoring system. In order to obtain the clutch service life data in a reasonable period of time, the concept of an accelerated life test (ALT) is applied in this study. For this purpose, a fully instrumented SAE#2 test setup designed and built by the industrial partner, Dana Spicer Off Highway Belgium, was made available.

According to the standard of the Society of Automotive Engineer (SAE) (*i.e.* SAE *J*2489) (SAE-International, 2012), an SAE#2 test setup is used to evaluate the friction characteristics of automatic transmission clutches with automotive transmission fluids (ATFs). It can also be used to conduct durability tests on wet friction clutch systems and to evaluate the performance variation as a function of the number of duty cycles. Normally, a typical SAE#2 test setup is equipped with a flywheel driven by an electric motor and the kinetic energy of this wheel is dissipated in a tested clutch (Ost et al., 2001).

An ALT can be realized by means of applying a higher mechanical energy to a tested clutch compared to the amount of energy transmitted by a clutch in normal operation. The energy level is normally adjusted by changing the initial relative velocity and/or the inertia of input and output flywheels. In this study, the ALTs were conducted on some wet friction clutches with different friction materials using a fully instrumented SAE#2 test setup. During the tests, all the clutches were lubricated with the same Automatic Transmission Fluid (ATF). The used SAE#2 test setup and the proposed ALT procedure are discussed in the following subsections.

3.1. SAE#2 test setup

The SAE#2 test setup used in the experiments, as depicted in Figure 4, consists of three main systems, namely: driveline, control and measurement system. The driveline comprises several components: an AC motor for driving the input shaft (1), an input velocity sensor (2), an input flywheel (3), a clutch pack (4), a torque sensor (5), output flywheel (6), an output velocity sensor (7), an AC motor for driving the output shaft (8), a hydraulic system (11-20) and a heat exchanger (21) for cooling the outlet ATF. An integrated control and measurement system (22) is used for controlling the ATF pressure (both for lubrication and actuation) to the clutch and for the initial velocity of both input and output flywheels as well as for measuring all relevant dynamic signals. It should be mentioned here that both velocity sensors are Hall-effect encoders sensing gears with the teeth number of 51. This means that, the resolution of the used rotational velocity sensors is 51 pulses per revolution.

3.2. Test specification

To experimentally verify the developed condition monitoring method for wet friction clutches in various conditions and configurations, a test scenario was designed. The general specification of the test scenario is given in Table 1. Five experiments were conducted in this study wherein a different clutch pack was used for each experiment. The energy applied to each clutch pack in the first four tests is set to a relatively high level; while the energy applied in the last test is set at a lower level, see Table 2. In terms of design, all the used clutch packs are identical, only the friction material

(a)

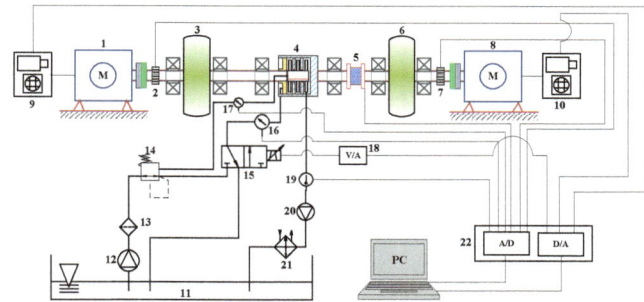

(b)

Figure 4. The SAE#2 test setup used in the study, (a) photograph and (b) scheme, courtesy of Dana Spicer Off Highway Belgium.

is different for each clutch pack, see Table 2. Lining materials of the friction discs used in all the tests are paper-based type while the materials of all the separator discs are steel. It should be noted that all the used friction discs, separator discs and ATF are commercial ones which can be found in the market. In all the tests, the inlet temperature and flow of the ATF were kept constant, see Table 1. Additionally, one can see in the table that the inertia of the input flywheel (drum-side) is lower than that of the output flywheel (hub-side).

Number of clutch packs to be tested	5
Number of friction discs in the clutch assembly	8
Inner diameter of friction disc (d_i) [mm]	115
Outer diameter of friction disc (d_o) [mm]	160
ATF	*John Deere J20C*
Lubrication flow [liter/minute]	18
Inlet temperature of ATF [°C]	85
Output flywheel inertia [kgm^2]	3.99
Input flywheel inertia [kgm^2]	3.38
Sampling frequency [kHz]	1

Table 1. General test specification.

3.3. Test procedure

Before an ALT is carried out to a wet friction clutch, a run-in test (lower energy level) is first conducted for 100 duty cycles

Clutch pack	Friction disc	Separator disc	Initial relative rotational velocity [rpm]
1	Dynax	Miba Tyzack	3,950
2	Raybestos I	Miba Tyzack	3950
3	Raybestos II	Miba Tyzack	3950
4	Wellman	Miba Tyzack	3950
5	Raybestos III	Miba Tyzack	2950

Table 2. ALT specifications.

in order to stabilize the contact surface. The run-in test procedure is in principle the same as the ALT procedure, but the initial relative rotational velocity of the run-in tests is lower than that of the ALTs. Figure 5 illustrates a duty cycle of the ALT that is carried out as follows. Initially, while both input flywheel (drum-side) and output flywheel (hub-side) are rotating at respective speeds in opposite direction, the two motors are powered-off and the pressurized ATF is simultaneously applied to a clutch pack at time instant t_f. The oil thus actuates the clutch piston, pushing the friction and separator discs towards each other. This occurs during the filling phase between the time instants t_f and t_a. While the applied pressure is increasing, contact is gradually established between the separator and friction discs which results in an increase of the transmitted torque and a simultaneous decrease of the relative velocity. Finally, the clutch is completely engaged when the relative velocity reaches zero at the lockup time instant t_l. As the inertia and the respective initial speed of the output flywheel (hub-side) are higher than those of the input flywheel, after t_l, both flywheels rotate together in the same direction as the output flywheel, see Figure 5. In order to prepare for the forthcoming duty cycle, both driving motors are braked at the time instant t_b, such that the driveline can stand still for a while.

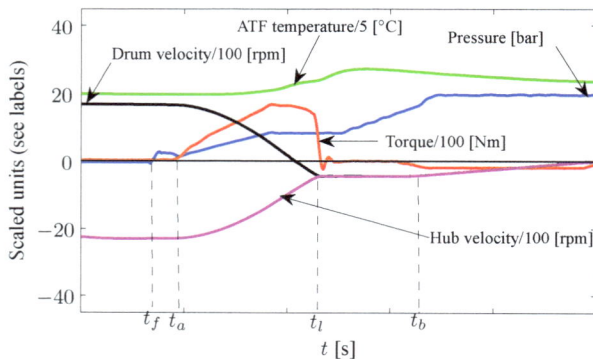

Figure 5. A representative duty cycle of wet friction clutches. Note that the transmitted torque drops to zero after the lockup time instant t_l because there is no external load applied during the test.

It is known that the normal load (pressure) has a significant

effect on the frictional characteristics. This means the pressure applied to wet friction clutches has also a significant effect on the engagement behavior which in turn influences the signal profile of the relative velocity. Since only the effect of the clutch degradation on the change of the relative velocity signal is of interest, the pressure signal applied to the clutch in each test is therefore kept the same.

The ALT procedure discussed above is continuously repeated until a given total number of duty cycles is attained. For the sake of time efficiency in measurement, all the ALTs are performed for 10000 duty cycles. Moreover, the ATF is continuously filtered, such that it is reasonable to assume that the used ATF has not degraded during all the tests.

Figure 6. Friction and separator discs after 10000 duty cycles, courtesy of Dana Spicer Off Highway Belgium.

4. RESULTS AND DISCUSSION

Figure 6 shows the photographs of friction and separator discs of a wet friction clutch after 10000 duty cycles, taken from the first clutch pack. From the figure it can be seen that the surfaces of the friction discs have become smooth and glossy. Nevertheless, it is evident that the separator discs are still in good condition.

Figure 7 shows the comparisons of the optical images and the surface profile of the friction material before and after the ALT. The images are captured using a **Zeiss microscope** and the surface profiles are measured along the sliding direction using a **Taylor Hobson Talysurf** profilometer. It can be seen in the figure that the surface of the friction material has become smooth and glossy and the clutch is therefore considered to have failed. The change of the color and the surface topography of the friction material is known as a result of the glazing phenomenon that is believed to be caused by a combination of adhesive wear and thermal degradation (Gao & Barber, 2002). Due to these two mechanisms, the surface pores of the friction material are blocked by the deposition of debris particles and/or the deposition of the ATF products.

Without loss of generality, energy density that is defined as the transmitted energy per unit of total contact area is introduced here for comparing the degradation rate of the tested clutches with different test conditions. The transmitted en-

(a)

(b)

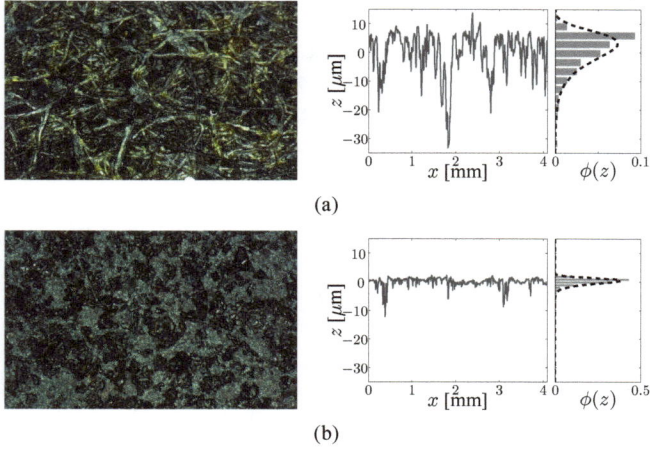

Figure 7. Comparison of the friction material before and after the ALT of 10000 duty cycles. (a) optical image (left) and the corresponding surface profile (right) of the friction material *before* the test, (b) optical image (left) and the corresponding surface profile (right) of the friction material *after* the test. Notice that z denotes the displacement of the profilometer stylus in Z-axis (perpendicular to the surface), x denotes the displacement of the profilometer stylus in X-axis (along the sliding direction) and $\phi(z)$ denotes the probability distribution function of the surface profile.

ergy E_{cycle} at a given duty cycle is computed as follows:

$$E_{cycle} = \int_{t_e}^{t_l} M\omega_{rel}dt, \qquad (14)$$

where M denotes the transmitted torque and $\omega_{rel} = \frac{2\pi n_{rel}}{60}$ denotes the relative velocity in rad/s. Hence, the energy density $\mathscr{E}_{cycle}$ transmitted by a wet friction clutch at a given duty cycle can be computed as follows:

$$\mathscr{E}_{cycle} = \frac{E_{cycle}}{N_f \mathscr{A}_f}, \qquad (15)$$

where N_f is the number of friction faces and $\mathscr{A}_f$ is the apparent contact area between friction disc and separator disc. By applying Equations (14) - (15) to the measured torque and relative velocity, the energy density of each test can be computed as presented in Table 3. As expected, it can be clearly seen from the table that the energy density $\mathscr{E}_{cycle}$ applied to the fifth clutch pack is lower than (approximately half of) that of other clutch packs.

The remainder of this section is structured as follows. First, the COFs of all the tested clutches are computed and their characteristics during the clutch service life are evaluated and discussed. Afterwards, the features proposed in this paper, *i.e.* the dimensionless engagement time and dimensionless dissimilarity measures, are extracted from the relative velocity signals. Finally, the proposed features are compared and evaluated with the mean COF which has been considered, as mentioned in the introductory section, as a *reference feature*.

Test	$\mathscr{E}_{cycle} \times 10^{-6}$ [Joule/m^2]
1	1.068
2	1.068
3	1.068
4	1.068
5	0.584

Table 3. The energy density $\mathscr{E}_{cycle}$ applied in the ALTs.

4.1. COF characteristic during the service life of the tested clutches

As previously stated in Section 1, the COF has been used for many years to monitor and evaluate the condition of wet friction clutches. Due to its strong correlation to the nature of clutch degradation, it is therefore reasonable to employ the COF as a reference feature for evaluating and justifying the relevance of the proposed features with respect to the progression of the clutch degradation. At a given duty cycle, the instantaneous COF μ of a wet friction clutch can be computed according to the following equation (Ost et al., 2001):

$$\mu = \frac{3M(r_o^2 - r_i^2)}{2N_f F_a (r_o^3 - r_i^3)}, \qquad (16)$$

where r_o is the outer radius of friction disc, r_i is the inner radius of friction disc, N_f is the number of friction faces and F_a is the axial force applied to the clutch. The applied force can be estimated based on the applied pressure p and the force of the returning spring F_s, *i.e.* $F_a \approx pA_p - F_s$, with A_p denoting the area of the piston.

Suppose that the spring force and friction disc geometry are given, the instantaneous COFs of all the tested clutch packs can be computed by applying Equation (16) to the experimental data (the measured torque and pressure signals). The spring force is determined here based on the deformation of the returning spring, when the piston and all the discs make contact, with respect to its rest position.

In order to quantify the global characteristic of the COF of a wet friction clutch, the mean COF μ_m as proposed in (Ost et al., 2001) can be applied for this purpose. For one duty cycle, this quantity is defined as follows:

$$\mu_m = \frac{1}{(t_l - t_e)} \int_{t_e}^{t_l} \mu d\tau. \qquad (17)$$

For a discrete data set, *i.e.* $\mu = [\mu_1, \mu_2, \ldots, \mu_j, \ldots, \mu_L]$, the mean COF computed with Equation (17) can be rewritten as follows:

$$\mu_m = \frac{1}{L} \sum_{j=1}^{j=L} \mu_j, \qquad (18)$$

with L denoting the dataset length of the instantaneous (discrete) COF μ_j and indexes $j = 1$ and $j = L$ correspond to the time instant t_e and t_l, respectively. For convenience, the

μ_m can be normalized according to the following equation:

$$\delta\hat{\mu}_m = \frac{\mu_m - \mu_m^r}{\mu_m^r}, \qquad (19)$$

with $\delta\hat{\mu}_m$ denoting the normalized mean COF and μ_m^r denoting the mean COF measured from the first duty cycle.

In Figure 8, the evolution of the mean COFs and the normalized ones are depicted. The dropping mean COFs can be explained as follows. As the degradation progresses, the surface of the friction material becomes smoother and debris particles are possibly entrapped on the surface pores of the friction material. Moreover, the deposition of ATF products may also blockade the surface pores of the friction material. This complex phenomenon is well known as a glazing process (Li et al., 2003). As a result of the glazing phenomenon, the ability of the ATF to escape from the approaching contact surfaces decreases. In this particular situation, the ATF stays between the contacting surfaces which hampers the occurrence of surface to surface contact corresponding to the *boundary lubrication* regime. Thus, the occurring friction is mainly controlled by the ATF resulting in lower COFs. In addition to this, the mechanical properties of the friction material change, *e.g.* reduction of the shear strength (Maeda & Murakami, 2003), which may have an additional effect on the COFs reduction. By considering that the ATF has not signif-

the tested friction materials have degraded to a certain extent. The level of friction material degradation is not only dependent on the amount of input energy, but also the design, *e.g.* material durability. As can be seen in Figure 8, the effect of the used friction material on the COF evolution can be observed. Despite the same energy level, the mean COF reduction of the *fourth* clutch pack with the Wellman friction material is *less* than that of the first three clutch packs. It can be seen in the figure that the dimensionless mean COF reduction after 10000 cycles of this particular friction material is approximately half of the others conducted at the same energy level (Dynax, Raybestos I and Raybestos II, see Table 2). Accordingly, one may conclude that the Wellman product used in the 4th ALT is more durable than the other friction materials tested in the study. In addition, the effect of energy level can also be observed from the data (compare the 5th ALT with the other tests). As expected, the lower the energy level the smaller the COF alteration will be.

4.2. Experimental verification of the developed condition monitoring method

The pressure signals obtained from the measurements are plotted in the left panels of Figures 9 - 13. The figures show that the pressure signals applied to a tested clutch during the ALT are relatively identical. This suggests that the effect of the pressure variation on the change of the relative velocity signal during the clutch lifetime can be assumed negligible.

Different from the fifth ALT, the service life relative velocity signals obtained from the first four ALTs show noticeable changes; see Figures 9 - 12. These noticeable changes can be explained as follows. In the first four tests, the initial velocity is set at high value which gives a large amount of energy to the clutches to be tested. A higher energy level applied on a clutch implies that a higher degradation rate occurs in the friction material. As a result, the frictional characteristics change significantly which is reflected by the noticeable change of the relative velocity signal. Accordingly, it can be deduced that with the same operational condition and identical clutches, the shape of the relative velocity signal profile of clutches is strongly dependent on its initial value, see Figures 11 and 13. As was expected, the higher the initial relative velocity is, the longer the engagement duration will be and vice versa.

In order to quantify the change of the relative velocity signals during the clutches service life, the proposed features as have been introduced in Section 2 are extracted from the service life data using Equations (10), (12) and (13). The plots of these features in function of the clutches service life are depicted in Figure 14. The figure clearly show the progress of clutch degradation which is embodied in the features evolution. This observation implies that the proposed features are relevant to monitor the progression of clutch degradation.

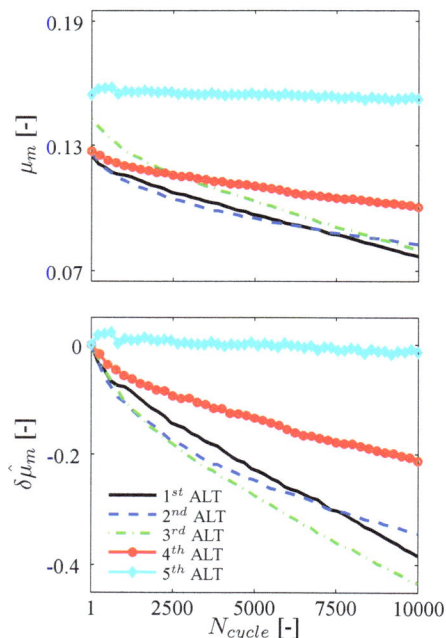

Figure 8. Evolution of the mean COF and its normalized value during the lifetime of the tested clutches.

icantly degraded during the tests (which is the case in this study), the progressive change of the COFs implies that all

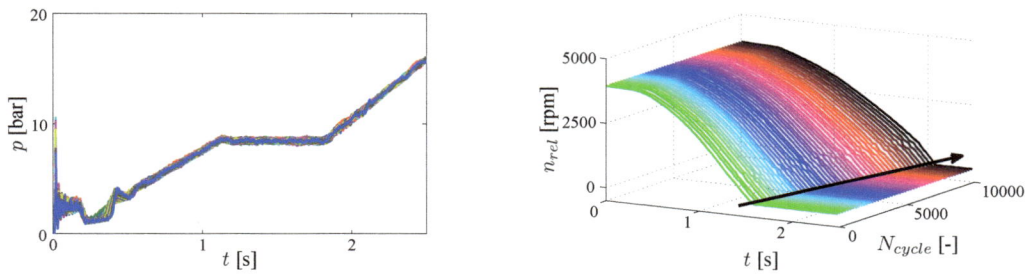

Figure 9. Pressure signals (left) and relative velocity signals (right) obtained from the *first* ALT.

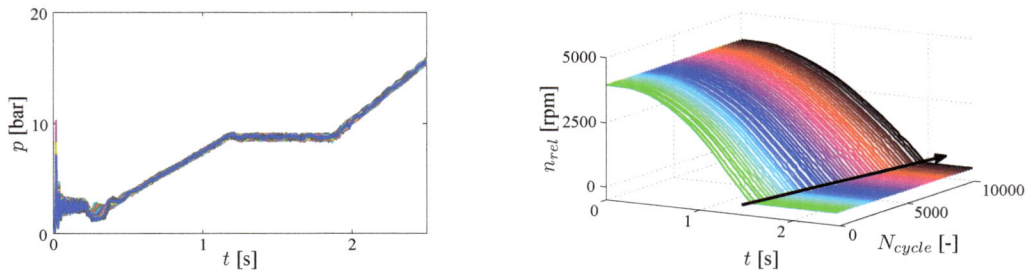

Figure 10. Pressure signals (left) and relative velocity signals (right) obtained from the *second* ALT.

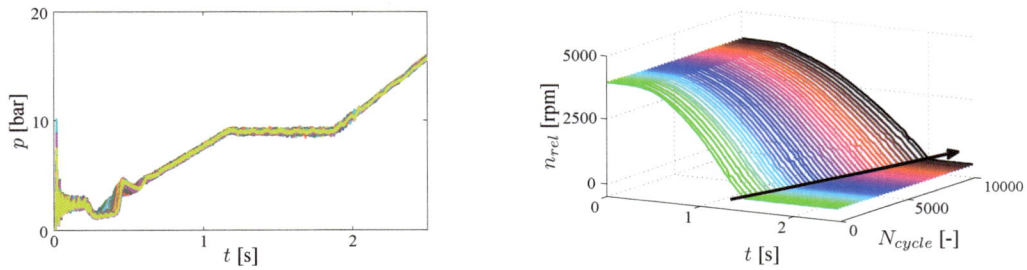

Figure 11. Pressure signals (left) and relative velocity signals (right) obtained from the *third* ALT.

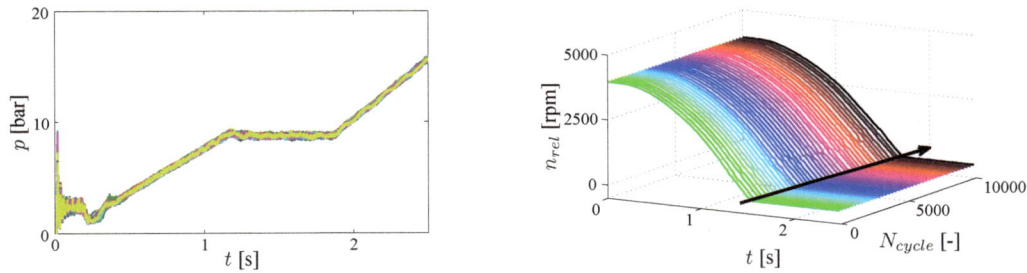

Figure 12. Pressure signals (left) and relative velocity signals (right) obtained from the *fourth* ALT.

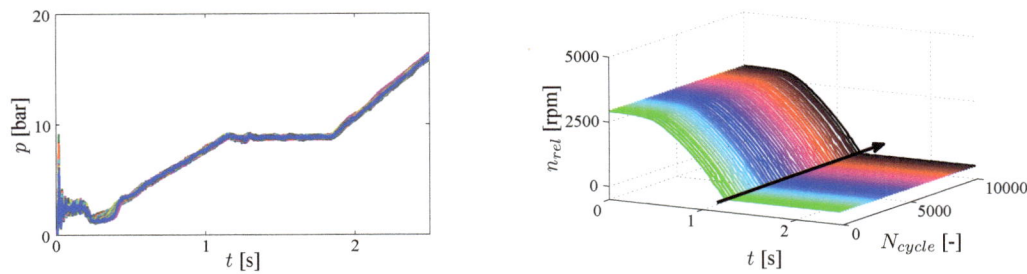

Figure 13. Pressure signals (left) and relative velocity signals (right) obtained from the *fifth* ALT.

The increasing dimensionless engagement duration as depicted in the upper panel of Figure 14 implies that the COF drops as the degradation progresses, see Figure 8, which confirms the experimental data available in the literature (Matsuo & Saeki, 1997; Ost et al., 2001; Maeda & Murakami, 2003; Fei et al., 2008). For the same operating conditions (*e.g.* the same applied pressure to the piston clutch), as has been mentioned previously, the dropping COF consequently lowers the transmitted torque during the clutch service life. When the transmitted torque becomes lower, the relative velocity decreases more slowly from its initial value to zero value.

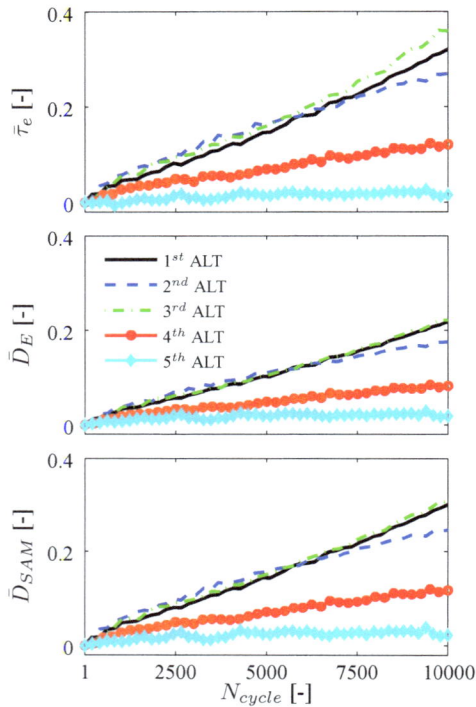

Figure 14. Evolution of the proposed dimensionless features during the lifetime of the tested clutches.

Since the relative velocity signal of interest obtained from the first duty cycle is taken as the *baseline* as discussed previously in Section 2, the dissimilarity measures at the first cycle are zero as can be seen in the middle and lower panels of Figure 14. When the clutch degradation progresses further, the shape of the relative velocity signal of interest deviates from the *baseline* which consequently results in an increase of the dissimilarity measures.

Remarkably, all the plots of the proposed features in function of the clutch duty cycles show linearly increasing trends with relatively small variations. The possible explanation for this observations is that all the tests were repeated with the same operational conditions. In practice, the operating conditions can vary and the energy level per duty cycle is typically lower than that applied in this study. Hence, the trends of the proposed features are not necessarily linear and the variations are not necessarily small. Because of possibly varying operating conditions, the trends of the proposed features may exhibit non-linear behaviors with large variations in real situations. In addition to this, the rate of change of the feature values in real-situations is possibly lower than that observed on the ALT results of this study.

5. CONCLUSION AND FUTURE WORK

A condition monitoring method for wet friction clutches that can be used in real-life applications has been developed and discussed in this paper. The method is based on monitoring the change of the relative velocity signal measured between the input and output shaft of a clutch. Three dimensionless parameters which are easy to compute, namely the *normalized engagement duration*, the *normalized Euclidean distance* and the *Spectral Angle Mapper (SAM) distance* are proposed in this paper as features for monitoring the condition of wet friction clutches. The developed method suggests that the sensors typically available in automatic transmissions, namely pressure and velocity sensors, can be employed for extracting the proposed features.

Service life data obtained from accelerated life tests (ALTs), carried out using an SAE#2 test setup on five different paper-based wet friction clutches, are employed in order to evaluate and verify the potential and the relevance of the proposed features for clutch condition monitoring. All the plots of the proposed features in function of the service life of the tested clutches exhibit clear trends, which can be associated with the progress of the clutch degradation. The trends are well correlated with the mean coefficient of friction (COF) which can be considered here as a *reference feature*. Remarkably, all the trends exhibited by the features proposed in the paper are (*monotonically*) increasing with relatively small variations. For akin clutches, it has been observed that the evolution of the features extracted from the clutches' life data obtained from the ALTs with a higher energy level, shows a steeper trend (larger slope) compared to the one with lower energy level; where this observation is also consistent with the result of the mean COF. The slope difference leads to a conclusion that clutches tested with higher energy level experience a higher degradation rate than those with lower energy level. Since the proposed features are able to expose the progression of clutch degradation, their relevance for monitoring and assessing the condition of wet friction clutches are justified.

So far, the developed clutch monitoring method has been evaluated and verified on ALTs under controlled environment, where (i) the inlet ATF temperature is controlled at a constant value, (ii) the applied pressure variation is relatively small and (iii) the external load is fixed. In practice, the operating condition can vary that may affect the accuracy of the developed monitoring method. The effects of the operational parameter variations on the proposed features need to be further investigated. A profound understanding may allow us to model the feature variations such that an accurate monitoring system can be achieved.

A. APPENDIX

The Euclidean distance D_E between two signals of interest is defined similarly as the Euclidean distance between two vectors $\boldsymbol{X}$ and $\boldsymbol{Y}$ as follows:

$$D_E(\boldsymbol{X}, \boldsymbol{Y}) = \sqrt{\sum_{i=1}^{K} (x_i - y_i)^2}. \qquad (A.1)$$

If Equation (A.1) is imposed in the following form:

$$\bar{D}_E(\boldsymbol{X}, \boldsymbol{Y}) = \frac{D_E(\boldsymbol{X}, \boldsymbol{Y})}{x_1 \sqrt{K}}, \qquad (A.2)$$

then we have:

$$0 \le \bar{D}_E(\boldsymbol{X}, \boldsymbol{Y}) \le 1. \qquad (A.3)$$

Therefore $\bar{D}_E(\boldsymbol{X}, \boldsymbol{Y})$ can be seen as a normalized version of $D_E(\boldsymbol{X}, \boldsymbol{Y})$.

Proof. Let Equation (A.2) be expanded as follows:

$$\bar{D}_E(\boldsymbol{X}, \boldsymbol{Y}) = \frac{D_E(\boldsymbol{X}, \boldsymbol{Y})}{x_1 \sqrt{K}}, \qquad (A.4)$$

$$= \sqrt{\frac{1}{K} \sum_{i=1}^{K} \frac{(x_i - y_i)^2}{x_1^2}}, \qquad (A.5)$$

$$= \sqrt{\frac{1}{K} \sum_{i=1}^{K} \left(\frac{x_i}{x_1} - \frac{y_i}{x_1} \right)^2}. \qquad (A.6)$$

Assume that the following properties hold:

$$\begin{cases} 0 \le x_i \le x_1, & i = 1, \ldots, K, \\ 0 \le y_j \le x_1, & j = 1, \ldots, K. \end{cases} \qquad (A.7)$$

From which it follows that:

$$\begin{cases} 0 \le \dfrac{x_i}{x_1} = \varepsilon_i \le 1, & i = 1, \ldots, K \\ \\ 0 \le \dfrac{y_j}{x_1} = \eta_j \le 1, & j = 1, \ldots, K. \end{cases} \qquad (A.8)$$

The inequalities above can be solved as follows:

$$\begin{aligned} -1 &\le & \varepsilon_i - \eta_i & \le 1 \\ 0 &\le & (\varepsilon_i - \eta_i)^2 & \le 1 \\ 0 &\le & \sum_{i=1}^{K} (\varepsilon_i - \eta_i)^2 & \le K \\ 0 &\le & \frac{1}{K} \sum_{i=1}^{K} (\varepsilon_i - \eta_i)^2 & \le 1 \\ 0 &\le & \sqrt{\frac{1}{K} \sum_{i=1}^{K} (\varepsilon_i - \eta_i)^2} \le 1, \end{aligned} \qquad (A.9)$$

which proves that:

$$0 \le \bar{D}_E(\boldsymbol{X}, \boldsymbol{Y}) \le 1. \qquad (A.10)$$

$\square$

ACKNOWLEDGMENT

All the authors wish to thank Dr. Mark Versteyhe of Dana Spicer Off Highway Belgium for the experimental support.

NOMENCLATURE

t	time
n_{rel}	relative velocity in rpm
ω_{rel}	relative velocity in rad/s
p	pressure
M	torque
F_a	axial force
E_{cycle}	energy transmitted for a given cycle
$\mathscr{E}_{cycle}$	energy density transmitted for a given cycle
$\mathscr{A}_f$	apparent contact area
A_p	contact area of piston
x	displacement of the profilometer stylus in X-axis along sliding direction
z	displacement of the profilometer stylus in Z-axis perpendicular to the surface
$\phi(z)$	probability distribution function of the surface profile
r_i	inner radius of friction disc
r_o	outer radius of friction disc
μ	instantaneous coefficient of friction
μ_m	averaged coefficient of friction
t_f	true reference time instant
t_l	true lockup time instant
$\hat{t}_f$	estimated reference time instant
$\hat{t}_l$	estimated lockup time instant
j, k	indices
n_n^f	floor noise relative velocity signal
n_p^f	floor noise pressure signal
τ	time record length
τ_e	engagement duration
$\boldsymbol{X}$	vector denoting a discrete relative velocity signal measured in an initial (healthy) condition
$\boldsymbol{Y}$	vector denoting a discrete relative velocity signal measured in an arbitrary condition
$\bar{\tau}_e$	normalized engagement duration
$\bar{D}_E$	normalized Euclidean distance
$\bar{D}_{SAM}$	normalized SAM distance
N_{cycle}	number of duty (engagement) cycles
N_f	number of friction faces

REFERENCES

Fei, J., Li, H.-J., Qi, L.-H., Fu, Y.-W., & Li, X.-T. (2008). Carbon-Fiber Reinforced Paper-Based Friction Material: Study on Friction Stability as a Function of Operating Variables. *Journal of Tribology*, *130*(4), 041605.

Gao, H., & Barber, G. C. (2002). Microcontact Model for Paper-Based Wet Friction Materials. *Journal of Tribology*, *124*(2), 414 - 419.

Guan, J., Willermet, P., Carter, R., & Melotik., D. (1998). Interaction Between ATFs and Friction Material for Modulated Torque Converter Clutches. *SAE Technical Paper*, *981098*, 245252.

Jullien, A., Meurisse, M., & Berthier, Y. (1996). Determination of tribological history and wear through visualisation in lubricated contacts using a carbon-based composite. *Wear*, *194*(1 - 2), 116 - 125.

Kruse, F., Lefkoff, A., Boardman, J., Heidebrecht, K., Shapiro, A., Barloon, P., et al. (1993). The spectral image processing system (SIPS) - interactive visualization and analysis of imaging spectrometer data. *Remote Sensing of Environment*, *44*(2-3), 145 - 163.

Li, S., Devlin, M., Tersigni, S., Jao, T.-C., Yatsunami, K., & Cameron., T. (2003). Fundamentals of Anti-Shudder Durability: Part I - Clutch Plate Study. *SAE Technical Paper*, *2003-01-1983*, 51 62.

Maeda, M., & Murakami, Y. (2003). Testing method and effect of ATF performance on degradation of wet friction materials. *SAE Technical Paper*, *2003-01-1982*, 45 - 50.

Matsuo, K., & Saeki, S. (1997). Study on the change of friction characteristics with use in the wet clutch of automatic transmission. *SAE Technical Paper*, *972928*, 93 - 98.

Nyman, P., Maki, R., Olsson, R., & Ganemi, B. (2006). Influence of surface topography on friction characteristics in wet clutch applications. *Wear*, *261*(1), 46 - 52. (Papers presented at the 11th Nordic Symposium on Tribology, NORDTRIB 2004)

Ompusunggu, A. P., Janssens, T., Al-Bender, F., Sas, P., & VanBrussel, H. (2011). Engagement behavior of degrading wet friction clutches. In *2011 IEEE/ASME International Conference on Advanced Intelligent Mechatronics (AIM2011)*.

Ompusunggu, A. P., Papy, J.-M., Vandenplas, S., Sas, P., & VanBrussel, H. (n.d.). A novel monitoring method of wet friction clutches based on the post-lockup torsional vibration signal. *Mechanical Systems and Signal Processing, submitted after revision.*

Ompusunggu, A. P., Papy, J.-M., Vandenplas, S., Sas, P., & VanBrussel, H. (2009). Exponential data fitting for features extraction in condition monitoring of paper-based wet clutches. In C. Gentile, F. Benedettini, R. Brincker, & N. Moller (Eds.), *The Proceedings of the 3rd International Operational Modal Analysis Conference (IOMAC)* (Vol. 1, p. 323-330). Starrylink Editrice Brescia.

Ompusunggu, A. P., Sas, P., VanBrussel, H., Al-Bender, F., Papy, J.-M., & Vandenplas, S. (2010). Pre-filtered Hankel Total Least Squares method for condition monitoring of wet friction clutches. In *The Proceedings of the 7th International Conference on Condition Monitoring and Machinery Failure Prevention Technologies (CM-MFPT)*. Coxmor Publishing Company.

Ompusunggu, A. P., Sas, P., VanBrussel, H., Al-Bender, F., Papy, J.-M., & Vandenplas, S. (2011). Normal-mode vibration analysis for condition monitoring of wet friction clutches. In *Proceedings of the 24th International Congress on Condition Monitoring and Diagnostic Engineering Management (COMADEM)*.

Ompusunggu, A. P., Sas, P., VanBrussel, H., Al-Bender, F., & Vandenplas, S. (2010). Statistical feature extraction of pre-lockup torsional vibration signals for condition monitoring of wet friction clutches. In *Proceedings of ISMA2010 Including USD2010*.

Ost, W., Baets, P. D., & Degrieck, J. (2001). The tribological behaviour of paper friction plates for wet clutch application investigated on SAE # II and pin-on-disk test rigs. *Wear*, *249*(5-6), 361 - 371.

Paclik, P., & Duin, R. P. W. (2003). Dissimilarity-based classification of spectra: computational issues. *Real-Time Imaging*, *9*(4), 237 - 244.

SAE-International. (2012). *SAE No. 2 Friction Test Machine Durability Test* (Vol. SAE J2489).

BIOGRAPHIES

Agusmian Partogi Ompusunggu is a project engineer at Flanders' MECHATRONICS Technology Centre (FMTC), Belgium. His research focuses in condition monitoring, prognostics, vibration analysis and measurement and tribology. He earned his bachelor degree in mechanical engineering (B.Eng) in 2004 from Institut Teknologi Bandung (ITB), Indonesia and master degree in mechanical engineering (M.Eng) in 2006 from the same technological institute. He is currently pursuing his PhD degree in mechanical engineering at Katholieke Universiteit Leuven (K.U.Leuven) Belgium.

Jean-Michel Papy received a Master Degree in Signal, Image and Acoustics from Paul Sabatier University, Toulouse, France, in 2000 and a PhD degree in Electrical Engineering from the K.U. Leuven, Belgium in 2005. His doctoral work was about the detection of transient signals and exponential data modeling using linear and multi-linear algebra. After his PhD, he has been working as a project engineer at Flanders MECHATRONICS Technology Centre (FMTC), Belgium. His current research interests include modeling of mechanical systems and sensor fusion.

Steve Vandenplas is a program leader at Flanders' MECHATRONICS Technology Centre (FMTC), Belgium. He received his Master's Degree of Electrotechnical Engineer in 1996 from the Vrije Universiteit Brussel (VUB), Belgium. In 2001, he received a PhD in Applied Science and started to work as R&D Engineer at Agilent Technologies for one year. Thereafter, he decided to work as a Postdoctoral Fellow at the K.U. Leuven in the Department of Metallurgy and Materials Engineering, in the research group material performance and non-destructive testing (NDT). He has been working at Flanders' MECHATRONICS Technology Centre (FMTC) since 2005, where he is currently leading FMTC's research program on "Monitoring and Diagnostics". His main interests are on machine diagnostics and condition based maintenance (CBM).

Paul Sas is a full professor at the Department of Mechanical Engineering of Katholieke Universiteit Leuven (K.U.Leuven), Belgium. He received his master and doctoral degree in mechanical engineering from K.U.Leuven. His research interest comprise numerical and experimental techniques in vibro-acoustics, active noise and vibration control, noise control of machinery and vehicles, structural dynamics and vehicle dynamics. He is currently leading the noise and vibration research group of the Department of Mechanical Engineering at K.U.Leuven.

Hendrik Van Brussel is an emeritus professor at the Department of Mechanical Engineering of Katholieke Universiteit Leuven (K.U.Leuven), Belgium. He obtained the degree of Technical Engineer in mechanical engineering from the Hoger Technisch Instituut in Ostend, Belgium in 1965 and an engineering degree in electrical engineering at M.Sc level from K.U.Leuven. In 1971 he got his PhD degree in mechanical engineering, also from K.U.Leuven. From 1971 until 1973 he was establishing a Metal Industries Development Center in Bandung, Indonesia and he was an associate professor at Institut Teknologi Bandung (ITB), Indonesia. He was a pioneer in robotics research in Europe and an active promoter of the mechatronics idea as a new paradigm in machine design. He has published more than 200 papers on different aspects of robotics, mechatronics and flexible automation. His research interests shifted towards holonic manufacturing systems and precision engineering, including microrobotics. He is Fellow of SME and IEEE and in 1994 he received a honorary doctor degree from the 'Politehnica' University in Bucharest, Romania and from RWTH, Aachen, Germany. He is also a Member of the Royal Academy of Sciences, Literature and Fine Arts of Belgium and Active Member of CIRP (International Institution for Production Engineering Research).

Towards A Model-based Prognostics Methodology for Electrolytic Capacitors: A Case Study Based on Electrical Overstress Accelerated Aging

José R. Celaya[1], Chetan S. Kulkarni[2], Gautam Biswas[3], and Kai Goebel[4]

[1] *SGT Inc. NASA Ames Research Center, Moffett Field, CA, 94035, USA*
jose.r.celaya@nasa.gov

[2, 3] *Vanderbilt University, Nashville, TN, 37235, USA*
chetan.kulkarni@vanderbilt.edu
biswas@eecsmail.vuse.vanderbilt.edu

[5] *NASA Ames Research Center, Moffett Field, CA, 94035, USA*
kai.goebel@nasa.gov

ABSTRACT

This paper presents a model-driven methodology for predicting the remaining useful life of electrolytic capacitors. This methodology adopts a Kalman filter approach in conjunction with an empirical state-based degradation model to predict the degradation of capacitor parameters through the life of the capacitor. Electrolytic capacitors are important components of systems that range from power supplies on critical avionics equipment to power drivers for electro-mechanical actuators. These devices are known for their comparatively low reliability and given their critical role in the system, they are good candidates for component level prognostics and health management. Prognostics provides a way to assess remaining useful life of a capacitor based on its current state of health and its anticipated future usage and operational conditions. This paper proposes and empirical degradation model and discusses experimental results for an accelerated aging test performed on a set of identical capacitors subjected to electrical stress. The data forms the basis for developing the Kalman-filter based remaining life prediction algorithm.

1. INTRODUCTION

This paper proposes the use of a model based prognostics approach for electrolytic capacitors. Electrolytic capacitors are critical components in electronic systems in aeronautics applications and in other domains. This type of capacitors are known to have lower reliability than other electronic components that are used in power supplies of avionics equipment and electrical drivers of electro-mechanical actuators of aircraft control surfaces. The field of prognostics for electronics is concerned with the prediction of remaining useful life (RUL) of the components and systems. This notion of condition-based health assessment leverages the knowledge of the device physics in order to model the degradation process, which is then used to predict remaining useful life as a function of current state of health and anticipated future operational and environmental conditions.

The prognostics methodology presented here, is based on the Bayesian tracking framework and a dynamic degradation model developed empirically from electrical overstress accelerated aging tests. A validation methodology is presented to assess the validity of the method using available run-to-failure data. The novelty of the approach consists on its ability to periodically estimate remaining useful life. This estimation process is condition-based in the sense that periodic measurements of the component under consideration are used in the estimation process. The contributions of this work are a dynamic degradation model and a model-based prognostics methodology for electrolytic capacitors. We present results for estimation of remaining useful life for five test cases. Predictions are made at several points in time during the life of the capacitors. Performance metrics like median relative accuracy and the α-λ metric demonstrate the effectiveness of our approach.

1.1. Motivation

The development of prognostics methodologies for electronic systems has become more important as more electrical systems are being used to replace traditional systems in several applications in the aeronautics, maritime, and automotive fields. The development of prognostics methods for electronics presents several challenges due to the great variety of components used in a system, the continual evolution of new electronics technologies, and a general lack of understanding of how electronics fail. Traditional reliability techniques in electronics tend to focus on understanding the time to failure for a batch of components of the same type, by running multiple experiments and making statistical estimates from the accumulated time to failure data. Recently, there has been a push to understand, in more depth, how faults progress as a function of time, loading, and environmental conditions. Furthermore, just until recently, it was believed that there were no precursor to failure indications for electronic components and systems. That is now understood to be incorrect, since electronic systems, much like mechanical systems, undergo a measurable wear process from which one can derive features that can be used to provide early warnings to failure. The indications of degradation caused by the wear can be detected fairly early, and by modeling the process, one can potentially predict the remaining useful life as a function of future use and environmental conditions.

Avionic systems perform critical functions on aircraft, greatly escalating the ramification of an in-flight malfunction (Bhatti & Ochieng, 2007; Kulkarni et al., 2009). Avionic systems combine physical processes, computational hardware and software. These elements present unique challenges for fault detection and isolation. A systematic analysis of these elements and their interaction is very important for the assessment of aircraft safety and to avoid catastrophic failures during flight.

Power supplies are critical components of modern avionic systems. In navigation systems, for instance, degradation and faults in the DC-DC converter power source unit propagate to the global positioning system (GPS) and other navigation subsystems, affecting the overall operations of the aircraft. Capacitors and power metal oxide field effect transistors (MOSFETs) are the two major components that cause performance degradation and failures in DC-DC converters (Kulkarni, Biswas, Bharadwaj, & Kim, 2010). Some of the more prevalent fault effects, such as a ripple voltage surge at the power supply output, can cause glitches in the GPS position and velocity output, and this in turn, if not corrected, can propagate and distort the navigation process.

Electrical motors are an essential element in electromechanical actuators systems that are being used to replace hydro-mechanical actuation in control surfaces of future generation aircrafts. Capacitors are used as filtering elements on power electronic systems, particularly for motor drivers. Electrical power drivers for motors require capacitors to filter the rail voltage for the H-bridges that provide bidirectional current flow to the windings of electrical motors. These capacitors help to ensure that the heavy dynamic loads generated by the motors do not perturb the upstream power distribution system. A failure in a rail voltage filter capacitor will have effects on the power distribution system and on performance of the motor, which will have cascading effects on the actuation process.

Low reliability and their criticality in avionics systems make electrolytic capacitors important candidates for focused health management solutions. In addition to this, degradation at component level could lead to cascading faults at subsystem and system levels. In order to mitigate the effects of capacitor failures in critical to safety systems, we introduce here, a condition-based prognostics methodology. This methodology develops the ability to identify degradation effects and to estimate the remaining life of the components progressively in time. This method will further allow for prognostics-based decision making for condition-based maintenance scheduling of the system or for implementation of mitigation strategies in case of contingencies during operation. In the next section we discuss briefly earlier work on capacitors at both component and system level.

1.2. Previous work

In earlier work, we studied the degradation of capacitors under nominal operation (Kulkarni, Biswas, Koutsoukos, Goebel, & Celaya, 2010b). The capacitors were components in a DC-DC converter, and their degradation was monitored every 100-120 hours of operation. Data was collected to determine the change in equivalent series resistance (ESR) and capacitance. An Arrhenius inspired ESR degradation model for time to failure computation was presented in Kulkarni, Biswas, Koutsoukos, Goebel, and Celaya (2010a). The data collected during the monitoring steps were used to compute the parameters of the model as well as for model validation.

In subsequent experimental work, we studied accelerated degradation in capacitors subjected to high charge/discharge cycles at a constant frequency (Kulkarni, Biswas, Koutsoukos, Celaya, & Goebel, 2010). A preliminary approach to computing RUL of electrolytic capacitors was presented in Celaya et al. (2011b).

In Kulkarni, Celaya, Goebel, and Biswas (2012b) we studied capacitor degradation under thermal overstress conditions. A physics-based degradation model, based on physics' first principles, was derived for the electrolytic capacitors and the derived model was employed in making RUL estimations based on a Bayesian tracking methodology (Kulkarni, Celaya, Goebel, & Biswas, 2012c, 2012a).

This paper here builds upon the initial studies presented in Celaya et al. (2011a) and Celaya et al. (2012).

1.3. Other related work and current art in capacitor prognostics

The output filter capacitor has been identified as one of the elements of a switched mode power supply that fails more frequently and has a critical impact on performance (Goodman et al., 2007; Judkins et al., 2007; Orsagh et al., 2005). A prognostics and health management approach for power supplies of avionics systems is presented in Orsagh et al. (2005). Results from accelerated aging of the complete supply have been discussed in terms of output capacitor and power MOSFET failures; but there is no modeling of the degradation process or RUL prediction for the power supply. Other approaches for prognostics for switched mode power supplies are presented in Goodman et al. (2007) and Judkins et al. (2007). The output ripple voltage and leakage current are presented as a function of time and degradation of the capacitor, but no details of the degradation process modeling, fault detection, and RUL prediction algorithms were presented.

A health management approach for multilayer ceramic capacitors is presented in Nie et al. (2007). This approach focuses on the temperature-humidity bias accelerated test to replicate failures. A method based on Mahalanobis distance is used to detect abnormalities in the test data; there is no prediction of RUL. A data driven prognostics algorithm for multilayer ceramic capacitors is presented in Gu et al. (2008). This method uses data from accelerated aging test to detect potential failures and to make an estimation of time of failure occurrence.

The approaches discussed above address fault detection and diagnostics methods using data-driven approaches. Our work focuses in prognostics, which is the natural progression from diagnostics. In addition, our methodology is based on dynamic degradation models used as part of a model-based prognostics framework.

2. PROGNOSTICS METHODOLOGY

A model-based prognostics methodology for electrolytic capacitors is presented in this section. This methodology relies on accelerated aging experiments to identify degradation behavior and to create time dependent degradation models. The process followed in the proposed methodology is presented in the block diagram in Figure 1 and described next.

Accelerated Aging: The methodology is based on results from an accelerated life test on real electrolytic capacitors. This test applies electrical overstress to commercial, off the shelf capacitors, in order to observe and record the degradation process and identify performance conditions in the neighborhood of the failure criteria in a considerably reduced time frame. A total of 6 test devices are used for this accelerated aging study. Electrochemical-impedance spectroscopy (EIS) is used periodically during the accelerated ag-

ing test to characterize the frequency response of the capacitor's impedance. Several measurements are made through the aging process, starting from measurements made under pristine condition all the way through to complete failure.

System Identification: A lumped-parameter model ($\mathcal{M}_1$) of the non-ideal capacitor impedance is assumed. This impedance model includes a capacitance element and a parasitic equivalent series resistance element. The EIS measurements along with the impedance model structure are used in a systems identification setting to estimate the model parameters. This is done for all the EIS measurements at different points in time during the aging experiment, resulting in time-dependent capacitance and ESR measurement trajectories reflecting capacitor degradation.

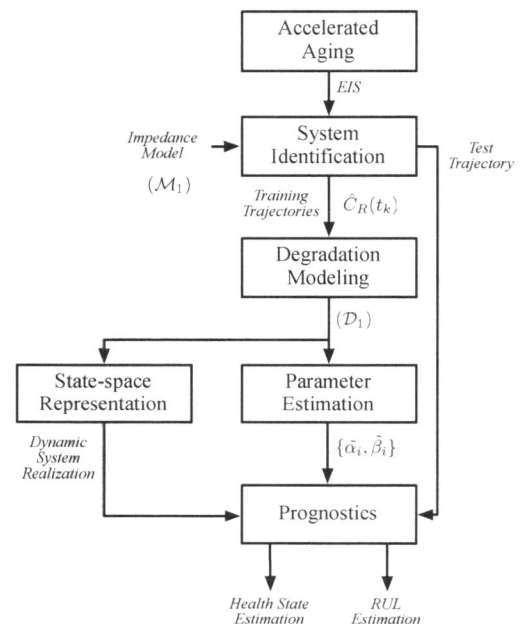

Figure 1. Methodology for the development of the capacitor prognostics approach.

Degradation Modeling: We present here an empirical degradation model that is based on the observed degradation process during the accelerated life test ($\hat{C}_R(t_k)$). The objective of the modeling is to generate a parametrized model ($\mathcal{D}_1$) of the time-dependent capacitance degradation as generated by the system identification step. A similar degradation model can be generated for ESR but it is not discussed in this work.

Parameter Estimation: A traditional training+test method is used to derive and validate the capacitor degradation model. The parameters of the degradation model are estimated using nonlinear least-squares regression. The quality of the parameter estimation results is satisfactory as to assume the estimated parameters ($\tilde{\alpha}_i, \tilde{\beta}_i$) are static (not time-dependent) during the prognostics process described next.

Prognostics: A Bayesian framework is employed to estimate (track) the state of health of the capacitor based on measurement updates for key capacitor parameters. The Kalman filter algorithm is used to track the state of health and the degradation model is used to make health state forecasts to be used in computation of remaining useful life once no further measurements are available.

3. ACCELERATED AGING EXPERIMENTS

Accelerated life test methods are often used in prognostics research as a way to assess the effects of the degradation process through time. It also allows for the identification and study of different failure mechanisms and their relationships with different observable signals and parameters. In this section we present a brief description of the accelerated electrical overstress experiment for studying capacitor degradation. We provide insights into the physical interpretation of the underlying degradation process under the electrical stress condition. Finally, we present a discussion on how the systems identification approach is used to compute the precursor to failure features based electrochemical impedance spectroscopy measurements obtained during the aging experiment.

3.1. Experimental Setup

Since the objective of this experiment is the study of the effects of high voltage on degradation of the capacitors, the capacitors were subjected to high voltage stress through an external power source using custom developed hardware. The voltage overstress is applied to the capacitors as a square waveform in order to subject the capacitor to continuous charge and discharge cycles.

At the beginning of the accelerated aging, the capacitors charge and discharge simultaneously; as time progresses and the capacitors degrade at different rates, the charge and discharge times vary for each capacitor. Even though all the capacitors under test are subjected to the same loading and operating conditions, their ESR and capacitance values change differently. We therefore monitor charging and discharging of each capacitor under test and measure the input and output voltages of the capacitor as well as the load current. Figure 2 illustrates the electrical overstress experiment's electrical circuit. A function generator is used to generate a square waveform, which is then amplified to the desired amplitude to be applied to the unit under test (UUT). A resistive load is used in series with the capacitor in order to emulate the loading side of a first order passive filter. Additional details on the accelerated aging system are presented in (Kulkarni, Biswas, Koutsoukos, Celaya, & Goebel, 2010).

For the experiment reported in this paper, a set of six capacitors was considered. Electrolytic capacitors of $2200\mu F$ capacitance, with a maximum rated voltage of $10V$, maxi-

mum current rating of $1A$ and maximum operating temperature of $105°C$ were used for the study. The electrolytic capacitors under test were characterized by EIS measurements before the start of the experiment and at different stages during the experiment execution. The experiment was conducted at room temperature.

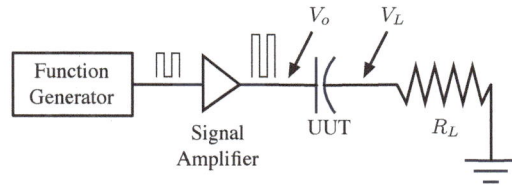

Figure 2. Block diagram of the experimental setup.

The EIS measurements were recorded every 8-10 hours of the total 180 plus hours of accelerated aging time in order to capture the degradation phenomenon in the ESR and capacitance values. During each measurement, the voltage source was shut down, the capacitors were discharged completely and then the EIS characterization procedure was carried out. This was done for all the six capacitors under test. A picture of the actual experiment setup is presented in Figure 3. For further details regarding the aging experiment, results and analysis of the measured data refer to (Kulkarni, Biswas, Koutsoukos, Celaya, & Goebel, 2010; Celaya et al., 2011b).

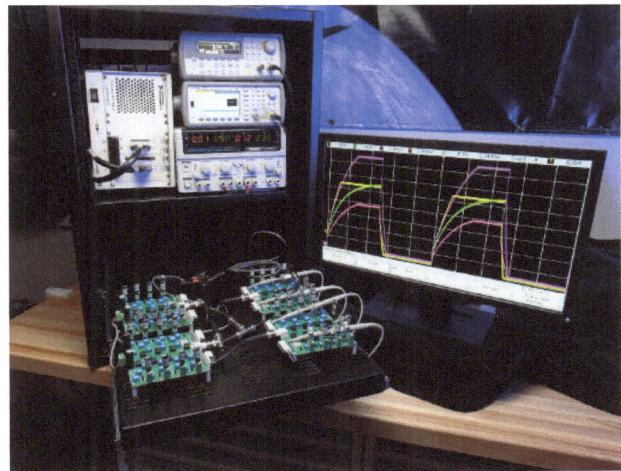

Figure 3. Electrical overstress aging experiment.

3.2. Physical interpretation of the degradation process

There are several factors that cause electrolytic capacitors to fail. Continuous degradation, in the form of gradual loss of functionality over a period of time, eventually results in the failure of the component. A complete loss of function is termed a *catastrophic* failure. Typically, this results in a short or open circuit in the capacitor.

In this work, we study the degradation of electrolytic capacitors operating under electrical stress, where $V_{applied} \geq V_{rated}$. During the charge/discharge process, the capacitors degrade over time. A study of the literature indicates that the degradation can be primarily attributed to electrolyte evaporation, leakage current and increased internal pressure due to gas released during chemical reactions (*International Standard IEC 60384-4-1: Fixed capacitors for use in electronic equipment*, 2007; MIL-C-62F, 2008; Kulkarni, Biswas, Koutsoukos, Goebel, & Celaya, 2010a). Our primary focus in this work is the study of the degradation due to electrolyte evaporation.

Figure 4 shows the structure of an electrolytic capacitor in detail. An ideal capacitor, while charging, would offer no resistance to the flow of current at its leads. However, the electrolyte that fills the space between the plates and the electrodes produces a small series resistance known as ESR. The quantitative changes in the ESR and capacitance values typify the current health state of the device, and represent the two primary precursors to failure. ESR and capacitance values were calculated after characterizing the capacitors at regular intervals. The heat generated due to current flow through the capacitor's internal resistance, ESR, during the operation of the capacitor, causes an increase in the internal temperature which also increases the rate of electrolyte evaporation. Decrease in electrolyte volume due to evaporation leads to further increases in ESR and decrease in the effective oxide surface area, resulting in capacitance decrease. The literature on capacitor degradation shows a direct relationship between electrolyte decrease with increase in ESR and decrease in capacitance value of the capacitor (Kulkarni, Biswas, Koutsoukos, Goebel, & Celaya, 2010b).

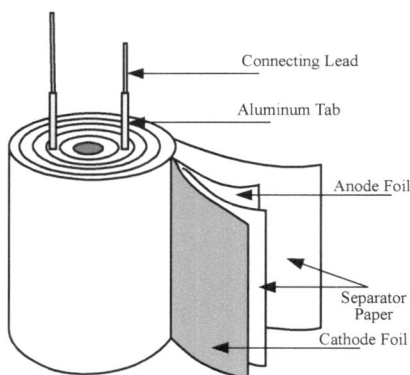

Figure 4. Electrolytic capacitor structure.

3.3. System identification for non-ideal capacitor model

ESR and capacitance values are estimated by using a system identification approach over a lumped parameter model ($\mathcal{M}_1$) consisting of the capacitance and the ESR in series as shown in Figure 5. It should be noted that the lumped-parameter model used to estimate ESR and capacitance offline, is not the degradation evolution model to be used on the online elements of the prognostics algorithm; it only allows us to estimate parameters, which provide indications of the degradation process through time. The impedance's frequency response is used for the system identification. Parameters such as ESR and capacitance are more difficult to estimate from the *in-situ* measurements of voltage and current partially available through the accelerated aging experiments. This can be done with a recursive estimation approach but it is not discussed in this paper.

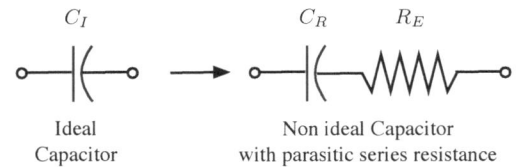

Figure 5. Lumped parameter model ($\mathcal{M}_1$) for a non-ideal capacitor.

The ESR and capacitance values were estimated from the capacitor electrochemical impedance frequency response measured using an SP-150 Biologic instrument. The ideal capacitor has complex impedance $Z_I = 1/sC_I$, where C_I is the ideal capacitance value. The complex impedance of model $\mathcal{M}_1$ is given by

$$\mathcal{M}_1 : Z = R_E + \frac{1}{sC_R}, \tag{1}$$

where R_E is the equivalent series resistance and C_R is the real capacitance.

Electrochemical impedance spectroscopy measurements are available to characterize the electrical performance of the capacitor throughout the aging experiment. Figure 6 shows Nyquist plots of the impedance measurements for capacitor #1 at pristine condition and after accelerated aging times 71, 161 and 194 hours. The degradation can be observed as the Nyquist plot shifts to the right as a function of aging time due to increase in R_E. These measurement are then used to estimate the parameters of the impedance model $\mathcal{M}_1$ from Eq. (1). The parameter estimation performed using the EIS instrument software (EC-Lab®). This is basically and optimization approach using an objective function defined as the aggregate of the squared error for all the frequencies where there is an impedance measurement from EIS. The error is computed based on the difference in magnitude of the model and the measured impedance. The optimization is set up to minimize the objective function by finding optimal values C_R^* and R_E^* for the $\mathcal{M}_1$ model.

This parameter estimation is performed every time an EIS measurement is made, resulting in values of C_R and R_E at different points in time through the aging of the components

($C_R(t_k)$ and $R_E(t_k)$). The average pristine condition ESR was measured to be $0.056\ m\Omega$ and the average pristine condition capacitance was measured to be $2123\ \mu F$ for the set of capacitors under test.

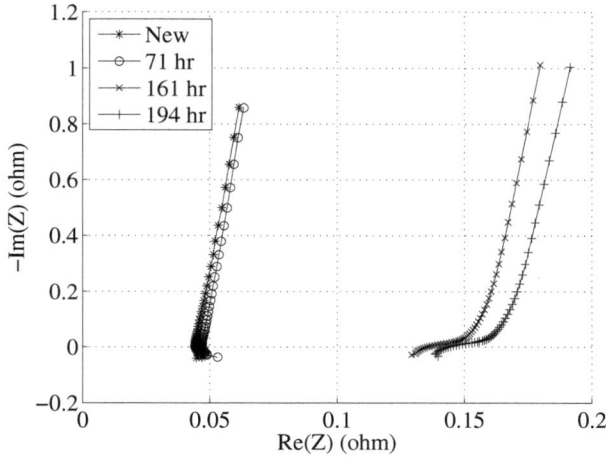

Figure 6. Electrochemical impedance measurements at different aging times.

Figure 7 shows the percentage increase in the ESR value for all the six capacitors under test over the aging time. Similarly, Figure 8 shows the percentage decrease in the value of the capacitance as the capacitor degrades over the aging time. Both parameters change through the aging experiment and are good candidates to be considered as precursor to failure features for the prognostics algorithm. As per standards MIL-C-62F (2008), a capacitor is considered unhealthy, if under electrical operation, its ESR increases by $280-300\%$ of its initial value; or the capacitance decreases by 20% below its pristine condition value. This information is used to set a crisp failure threshold for the RUL estimation process.

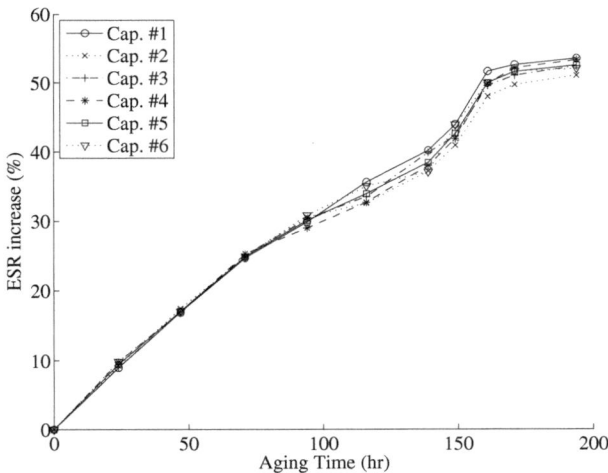

Figure 7. Percentage ESR increase as a function of time.

From the plots in Figure 7 we observe that for total experi-

ment time, the ESR value increased by $54\%-55\%$ for all the capacitors; while over the same period of time, the capacitance decreased by more than 20% (the threshold mark for a healthy capacitor) (see Figure 8). As a result, the percentage capacitance loss is selected as a precursor to failure variable to be used in the degradation model development presented next.

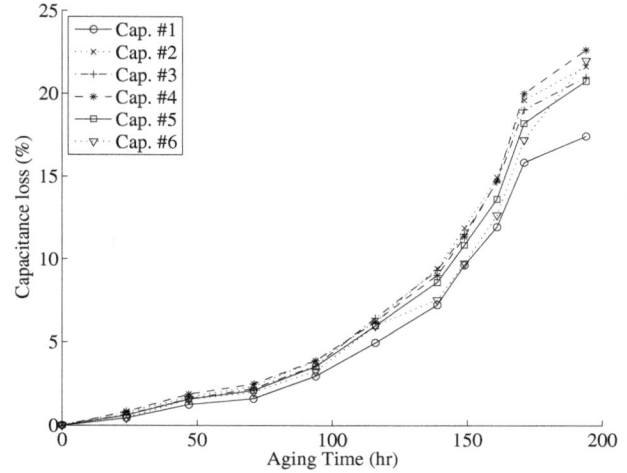

Figure 8. Percentage capacitance loss as a function of time.

4. DEGRADATION MODELING FOR PROGNOSTICS

This section presents the details of the degradation model development. A degradation model is an essential part of a model-based prognostics algorithm and it is typically application dependent. A model is formulated based on the empirical evidence of the time evolution of the degradation process from experiments presented in the previous section, particularly, capacitance loss as illustrated by Figure 8.

4.1. Nominal operation model

The non-ideal capacitor model $\mathcal{M}_1$ can be used as part of electronics circuits that make use of capacitors. An example is the low-pass filter implementation in Figure 9. In this circuit, input voltage V_i is considered as the voltage to be filtered and the voltage across the capacitor (this includes R_E as well) is the output voltage V_o that is filtered. Let $v(t) = V_o(t)$ and $u(t) = V_i(t)$ in the low-pass system circuit with non-ideal capacitor shown in Figure 9. A state-space realization model ($\mathcal{M}_2$) of the dynamic system is given by

$$\mathcal{M}_2: \begin{cases} \dot{z}(t) = \dfrac{-1}{C_R(R+R_E)}z + \dfrac{1}{C_R(R+R_E)}u(t), \\ v(t) = \left[1 - \dfrac{R_E}{R+R_E}\right]z + \dfrac{R_E}{R+R_E}, \end{cases}$$

(2)

where $z(t) = V_C(t)$ is the state variable representing the capacitor voltage, C_R, R_E and R are system parameters.

Furthermore, C_R and R_E are parameters that will change through time as the capacitor degrades, but are considered as static (constant) for the nominal operation model $\mathcal{M}_2$. Model $\mathcal{M}_2$ describes the nominal dynamics of a low-pass filter with a non-ideal capacitor. This model by itself is not sufficient to implement a model-based prognostics algorithm since the degradation process as reflected on model parameters is not modeled. Degradation models describing the time evolution of $R_E(t)$ or $C_R(t)$ are needed in order to enhance $\mathcal{M}_1$ or $\mathcal{M}_2$ for model-based prognostics. Nevertheless, $\mathcal{M}_1$ or $\mathcal{M}_2$ are useful in this form for model-based fault detection and isolation, which is not covered in this work.

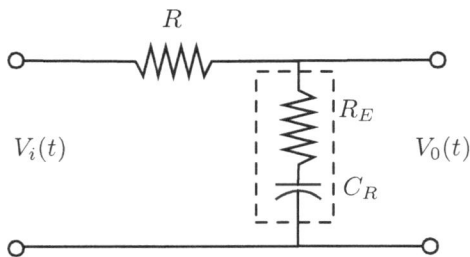

Figure 9. Circuit schematic for a low pass filter with non-ideal capacitor (model $\mathcal{M}_2$).

4.2. Degradation model

The percentage loss in capacitance is used as a precursor of failure variable and it is used to build a model of the degradation process. This model relates aging time to the percentage loss in capacitance. Let $C_l(t)$ be the percentage loss of capacitance due to degradation as shown by Figure 8. The following equation represents a dynamic first-order *degradation model* ($\mathcal{D}_1$) of the capacitance parameter in the non-ideal capacitor model $\mathcal{M}_1$.

$$\mathcal{D}_1 : C_l(t) = e^{\alpha t} + \beta, \qquad (3)$$

Here, α and β are degradation model parameters that will be estimated from the experimental data of accelerated aging experiments (degradation trajectories from Figure 8).

In order to estimate the model parameters, five capacitors are used for estimation, and the remaining capacitor is used to test the prognostics algorithm. This results in five leave-one-out cases for validation of the prognostics algorithm results. A nonlinear least-squares regression algorithm is used to estimate the model parameters α and β. Table 1 presents the definition of the test cases and the parameter estimation results. The parameter estimates ($\tilde{\alpha}, \tilde{\beta}$) and corresponding 95% confidence intervals are presented for parameters α and β. In addition, the error variance (σ_v^2) is included as a way to assess the quality of the estimation.

Figure 10 shows the estimation results for test case T_6. The experimental data are presented together with results from the exponential fit function. It can be observed from the residuals

that the estimation error increases with time. This is to be expected since the last data point measured for all the capacitors fall slightly off the exponential model. It should be noted that this degradation model with static parameters would be used in a Bayesian tracking framework as an online recursive estimation of $C_l(t)$. This will help to overcome the degradation model limitation to represent the behavior close to the failure threshold given the tracking framework ability to compensate the estimation as measurements become available.

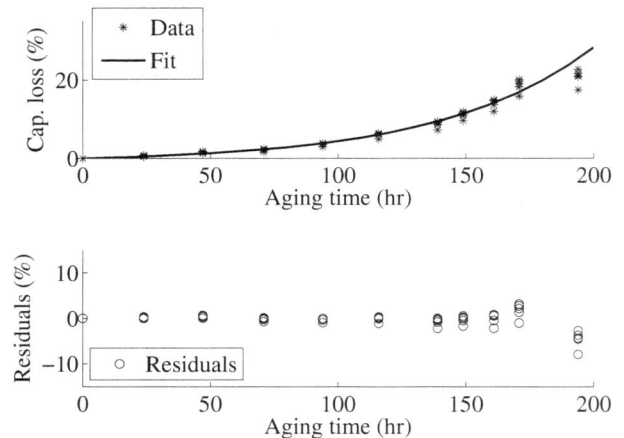

Figure 10. Estimation results for the empirical degradation model.

4.3. State-space realization for tracking

The estimated degradation model is used as part of a Bayesian tracking framework to be implemented using the Kalman filter technique. This method requires a state-space dynamic model relating the degradation level at time t_k to the degradation level at time t_{k-1}. The procedure to obtain a state-space model for $\mathcal{D}_1$ is as follows. The non-linear (in time evolution) exponential behavior described in the model is represented as a first order differential equation, which can represent the time evolution of $C_l(t)$. Then, the model is discretized in time in order to obtain a discrete-time state-space model $\mathcal{D}_2$. From Eq. (3) we have that $C_l(t) = e^{\alpha t} + \beta$, taking the first derivative with respect to time and substituting $e^{\alpha t} = C_l(t) - \beta$ from Eq. (3) we have

$$\dot{C}_l = \frac{dC_l(t)}{dt} = \alpha C_l(t) - \alpha \beta. \qquad (4)$$

Taking the finite difference approximation for $\dot{C}_l$ with time interval Δt we have

$$\frac{C_l(t) - C_l(t - \Delta t)}{\Delta t} = \alpha C_l(t - \Delta t) - \alpha \beta, \text{ and}$$

$$C_l(t) = (1 + \alpha \Delta_t) C_l(t - \Delta t) - \alpha \beta \Delta_t.$$

Letting $t_k = t$ and $t_{k-1} = t - \Delta t$ we get the state-space degradation model

$$\mathcal{D}_2 : C_l(t_k) = (1 + \alpha \Delta_k) C_l(t_{k-1}) - \alpha \beta \Delta_k. \qquad (5)$$

Validation test	Test capacitor	Training capacitors	$\tilde{\alpha}$ (95% CI)	$\tilde{\beta}$ (95% CI)	σ_v^2
T_2	#2	#1, #3–#6	**0.0162** (0.0160, 0.0164)	**-0.8398** (-1.1373, -0.5423)	1.8778
T_3	#3	#1, #2, #4–#6	**0.0162** (0.0160, 0.0164)	**-0.8287** (-1.1211, -0.5363)	1.9654
T_4	#4	#1–#3, #5, #6	**0.0161** (0.0159, 0.0162)	**-0.8217** (-1.1125, -0.5308)	1.8860
T_5	#5	#1–#4, #6	**0.0162** (0.0161, 0.0164)	**-0.7847** (-1.1134, -0.4560)	2.1041
T_6	#6	#1–#5	**0.0169** (0.0167, 0.0170)	**-1.0049** (-1.2646, -0.7453)	2.9812

Table 1. Degradation model ($\mathcal{D}_1$) parameter estimation results.

This model can be used in a Bayesian tracking framework in order to continuously estimate the value of the loss in capacitance through time as measurement become available.

5. Model-based Prognostics Framework

A model-based prognostics algorithm based on Kalman filter and a physics inspired empirical degradation model is presented. The methodology consists of the following three main steps and it is depicted in Figure 11.

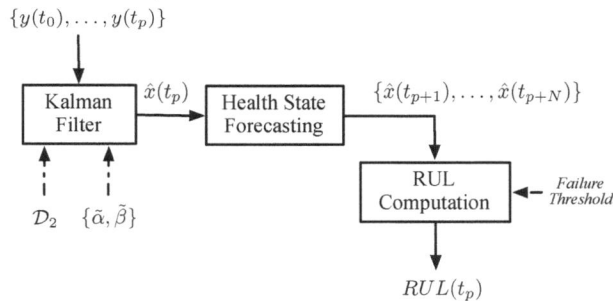

Figure 11. Model-based prognostics methodology

1. *State tracking (Kalman Filter)*: The capacitance loss $C_l(t)$ is defined as the state variable to be estimated and the degradation model is expressed as a discrete time dynamic model ($\mathcal{D}_2$) in order to estimate capacitance loss as new measurements become available. Direct measurements of the capacitance are available for the filtering process.

2. *Health state forecasting*: It is necessary to forecast the state variable once there are no more measurements available at time of RUL prediction t_p. This is done by evaluating the degradation model through time using the state estimate at time t_p ($\hat{x}(t_p)$) as initial value.

3. *Remaining life computation*: $RUL(t_p)$ is computed as the time difference between time of prediction t_p and the end-of-life (EOL) time. End of life is defined as the time at which the forecasted state crosses the failure threshold value.

This process is repeated for different values of t_p through the life of the component under consideration.

5.1. Kalman filter for state estimation

A state-space dynamic model is needed for the state estimation. The state variable x_k at time t_k is defined as the percentage capacitance loss $C_l(k)$. Since the system measurements are percentage loss in capacitance as well, the output equation is given by $y_k = hx_k$, where the value of h is equal to one. The following system structure is used in the implementation of the state estimation using the Kalman filter.

$$\begin{aligned} x_k &= A_k x_{k-1} + B_k u + v, \\ y_k &= hx_k + w, \end{aligned} \tag{6}$$

where,

$$\begin{aligned} A_k &= (1 + \Delta_k), \\ B_k &= -\alpha\beta\Delta_k, \\ h &= 1, \\ u &= 1. \end{aligned} \tag{7}$$

The time increment between measurements Δ_k is not constant since measurements were taken at non-uniform sampling rate. This implies that some of the parameters of the model in Eq. (6) will change through time. Furthermore, v and w are normal random variables with zero mean and Q and R variance respectively. The description of the Kalman filtering algorithm is omitted from this article. A thorough description of the algorithm can be found in Stengel (1994), a description of how the algorithm is used for forecasting can be found in Chatfield (2003) and an example of its usage for prognostics can be found in (Saha et al., 2009).

5.2. Future state forecasting

The use of the dynamic degradation model for health-state forecasting requires the time evolution of the state without updating the error covariance matrix and the posterior distribution of the state vector. Basically, the infrastructure provided by the Kalman filter for updating the state given new measurements on the system is not required in this step. The state equation from the discrete-time system in Eq. (6) is evaluated recursively n times ensuring the forecasted state $\hat{x}_{p+n}$ crosses the failure threshold. The noise variable v is also omitted. The n step ahead forecasting equation is given below by Eq. (8). The last update is done at the time of the last measurement t_l. Note that $t_l = t_p$ for this particular prognostics implementation. This is due to the authors' decision to make RUL predictions at each time a new measurement is available.

$$\hat{x}_{p+n} = A^n x_p + \sum_{i=0}^{n-1} A^i B \qquad (8)$$

The subscripts from parameters A and B are omitted since a constant Δ_t is used in the forecasting mode (one prediction every hour).

5.3. Remaining useful life computation

Computing the RUL based on the forecasting of the health states requires the identification of the EOL time based on the failure threshold. Defining $RUL(t_p)$ as the remaining useful life,

$$RUL(t_p) = t_{EOL} - t_p. \qquad (9)$$

The time at end-of-life (t_{EOL}) is a continuous variable which is computed from the forecast $\hat{x}_{p+n}$. Let $\hat{x}_{p+j}$ be the first forecast value to cross the failure threshold. A linear interpolation between $\hat{x}_{p+j}$ and $\hat{x}_{p+j-1}$ is used to compute t_{EOL}.

5.4. Noise models

The model noise variance Q was estimated from the model regression residuals for each test case presented in Table 1. The variance is listed in the last column Table 1. This variance was used for the model noise in the Kalman filter implementation. The measurement noise variance R is also required in the filter implementation. This variance was computed from the direct measurements of the capacitance with the EIS equipment, the observed variance is 4.99×10^{-7}.

6. PREDICTION OF REMAINING USEFUL LIFE RESULTS

State estimation and RUL prediction results are discussed for test case T_6. Figure 12 shows the result of the filter tracking the degradation signal. The residuals show an increased error with aging time. This is to be expected given the results observed from the model estimation process. It should be noted that the scale of the magnitude of the residuals is 10^{-7}. This is to be expected given the capability of the Bayesian tracking methods to make corrections to the state estimates based on direct or indirect measurements of the state variable being estimated. Even though the empirical degradation model with static parameters $\mathcal{D}_1$ does not perfectly represent the observed degradation data (Figure 10), the Kalman filter is able

to make the appropriate corrections resulting on a good online state estimation performance.

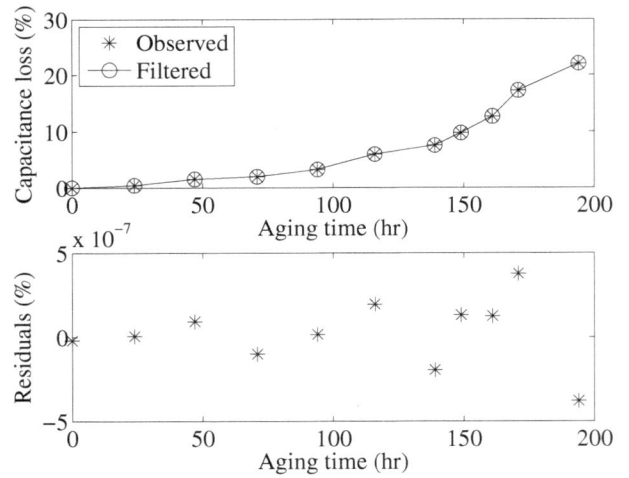

Figure 12. Tracking results for the Kalman filter implementation applied to test case T_6.

Figure 13 presents results from the remaining useful life prediction algorithm at time $t_p = 161$ (hr), which is the time at which ESR and C measurements are taken. The failure threshold is considered to be a crisp value of 20% decrease in capacitance. End of life is defined as the time at which the forecasted percentage capacity loss trajectory crosses the failure threshold. Therefore, RUL is t_{EOL} minus 161 hours. Figures 14 and 15 present the tracking and forecasting results for test case T_6. Appendix A presents the prediction plots for the remaining of the validation cases.

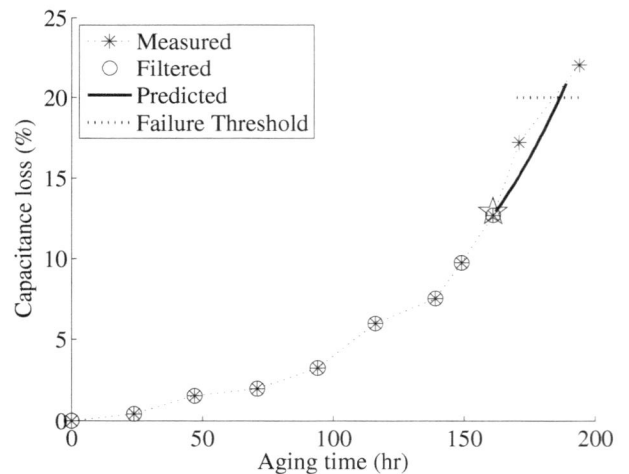

Figure 13. RUL prediction at time 161 (hr) for test case T_6.

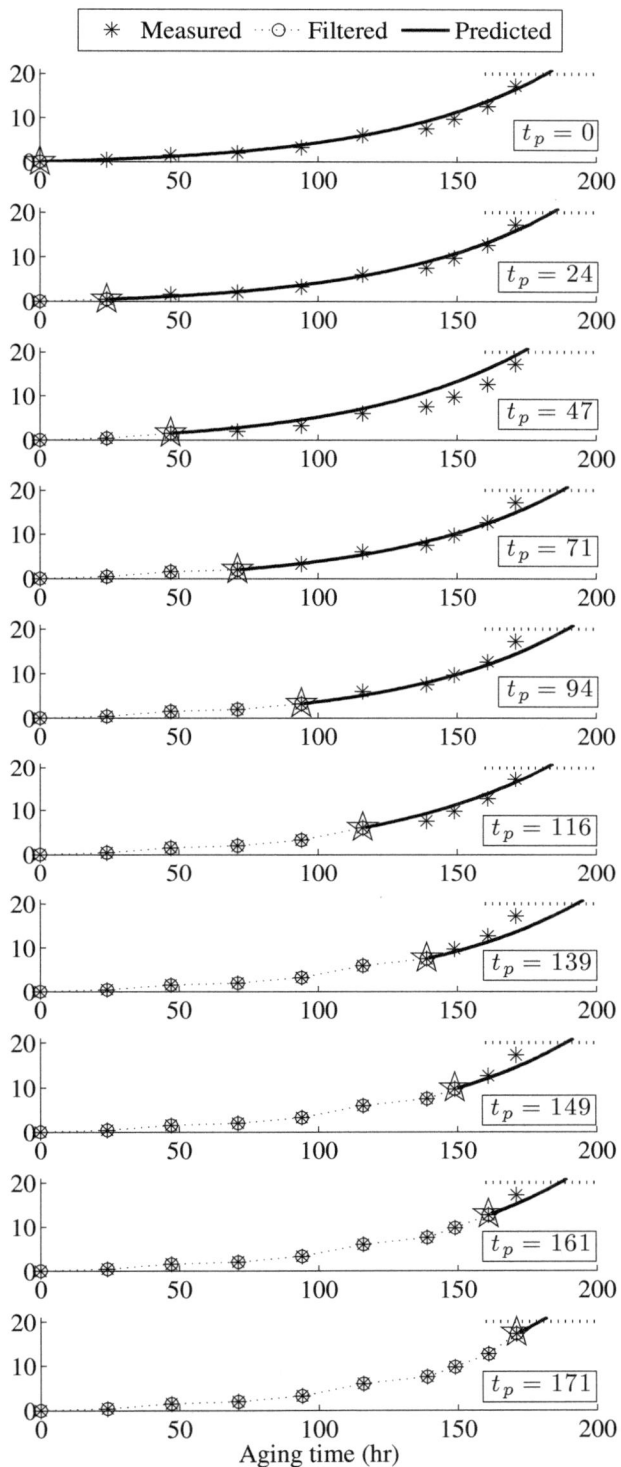

Figure 14. T_6: Health state estimation and forecasting of capacitance loss (%) at different times t_p during the aging time; $t_p = [0, 24, 47, 71, 94, 116, 139, 149, 161, 171]$.

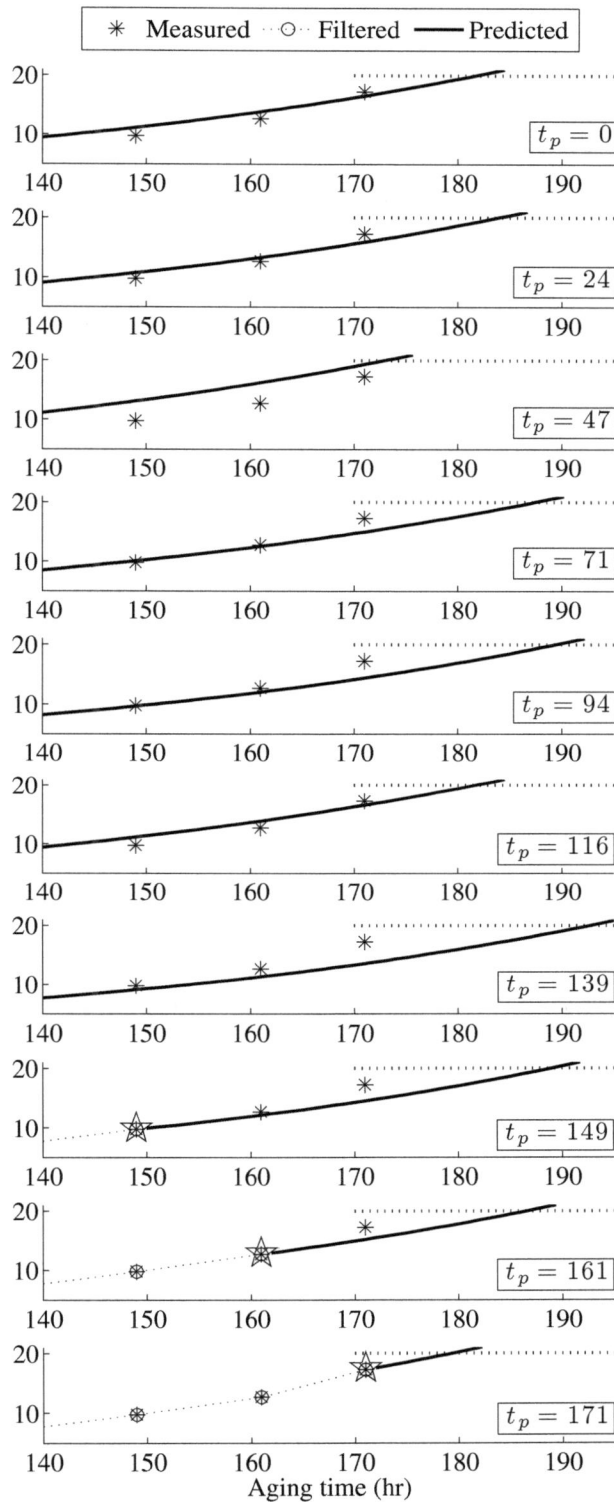

Figure 15. T_6: Detail of the health state estimation and forecasting of capacitance loss (%) at different times t_p during the aging time; $t_p = [0, 24, 47, 71, 94, 116, 139, 149, 161, 171]$.

Figure 14 illustrates the health-state estimation process and forecasting for the capacitance loss at different points during the aging time ($t_p = [0, 24, 47, 71, 94, 116, 139, 149, 161, 171]$ hours). The star symbol depicts the final state update (Kalman filter output) and the bold continuous line represents the state forecasted until the failure threshold is crossed. The failure threshold is depicted with a dotted horizontal line at 20%. RUL estimations are made after each point in which measurements are available. It can be observed that the predictions become better as the prediction is made closer to the actual EOL. This is possible because the estimation process has more information to update the estimates as it nears EOL. Figure 15 presents a zoomed-in version of Figure 14 focusing in the area close to the failure threshold.

An α-λ prognostics performance metric is presented in Figure 16 for validation test T_6. The continuous black line (RUL^*) represents ground truth and the shaded region corresponds to a 30% ($\alpha = 0.3$) error bound in the RUL prediction. This metric specifies that the prediction be within the error bound halfway between the first prediction and EOL ($\lambda = 0.5$). In addition, this metric allows us to visualize how the RUL prediction performance changes as data closer to EOL becomes available. Appendix B presents the α-λ metric plots for the remaining validation cases. Details on the prognostics performance metrics used in this work are available in Saxena, Celaya, Saha, Saha, and Goebel (2010).

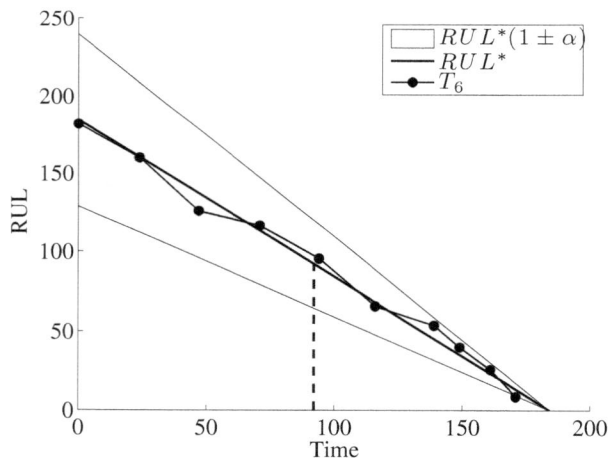

Figure 16. Performance based on α-λ performance metric.

6.1. Validation tests

Since the total number of capacitors used for the experiment was small (six), a leave-one out validation approach was used to assess the validity of the methodology. In this approach, the prognostics algorithm was executed in a single capacitor out of the set of six available. The remaining five capacitors are used for the parameter estimation of the degradation

model. Only five of the six available capacitors crossed the 20% capacitance loss failure threshold, as a result, only five validation cases were considered.

Table 2 shows performance based on the relative accuracy (RA) metric in Eq. (10). The RA metric allows for an assessment of the percentage accuracy relative to the ground-truth value. RA values of 100 represent perfect accuracy. The RA is presented for all the test cases for different prediction times. The last column of Table 2 represents the median RA of all the test cases for a particular prediction time. It is observed that the RA values decrease considerably for $t_p = 171$. This is consistent with previous observations indicating that the algorithm with a fixed-parameter model is not able to cope with the sudden jump in exponential behavior present around the 171 hour. This is a limitation that could be overcome by either an enhanced degradation model or a an online estimation of degradation model parameters using a more sophisticated Bayesian tracking method like extended Kalman filter or particle filter. It should be noted that the tracking algorithm has only ten measurement available to update the estimates, this has also an impact on the algorithm's ability to converge quickly to a small state estimation error. Having more measurements available through the aging experiments will have a positive impact on the prognostics performance.

$$RA = 100 \left(1 - \frac{\left| RUL^* - RUL' \right|}{RUL^*} \right) \qquad (10)$$

t_p	RA_{T2}	RA_{T3}	RA_{T4}	RA_{T5}	RA_{T6}	$\widetilde{RA}$
24	94.8	95.5	91.9	96.9	99.7	**95.5**
47	97.4	99.3	96.4	96.7	91.7	**96.7**
71	87.5	91.9	84.5	94.1	97.1	**91.9**
94	85.6	90	78.9	94.8	94.2	**90**
116	86	99.1	76.5	98	96.2	**96.2**
139	77.8	95.8	53.1	96.7	81.1	**81.1**
149	82.1	98.4	46.9	94.8	86.6	**86.6**
161	77.2	87.3	16.6	87.5	89.8	**87.3**
171	26.6	26.4	N/A	34.8	63.7	**30.7**

Table 2. Validation based on relative accuracy (RA) metric.

Table 3 summarizes results for the remaining life prediction (RUL') for all the test cases. The predictions at all points in time where measurements are available are included. The second column (RUL^*) indicates the RUL ground-truth use for computation of prediction errors. The results of the methodology are presented visually with the tracking and forecasting plots, the $\alpha - \lambda$ metric. The tabular results are included to provide needed information to compute other prognostics performance metrics and for algorithm performance comparison under the same run to failure dataset.

t_p	RUL^*	RUL'_{T2}	RUL'_{T3}	RUL'_{T4}	RUL'_{T5}	RUL'_{T6}
24	151.04	158.84	164.88	158.76	167.76	159.89
47	128.04	131.32	134.08	128.35	135.32	125.91
71	104.04	117.01	119.88	115.37	122.63	116.41
94	81.04	92.69	96.64	93.09	97.6	95.42
116	59.04	67.28	65.39	67.77	69.5	65.71
139	36.04	44.01	44.72	46.88	49.4	53.75
149	26.04	30.67	32.41	33.55	35.92	39.95
161	14.04	17.23	18.28	18.2	22.64	25.6
171	4.04	1.07	2.89	N/A	5.52	8.45

Table 3. Summary of RUL forecasting results for all test cases.

7. CONCLUSION

This paper presents a RUL prediction algorithm based on accelerated life test data and an empirical degradation model. The main contributions of this work are:

1. the identification of the lumped-parameter model (Figure 5) for a real capacitor as a viable reduced-order model for prognostics-algorithm development;

2. the identification of the ESR and C model parameters as precursor of failure features;

3. the development of an empirical degradation model based on accelerated life test data which accounts for shifts in capacitance as a function of time;

4. the implementation of a Bayesian based health state tracking and remaining useful life prediction algorithm based on the Kalman filtering framework.

One major contribution of this work is the prediction of remaining useful life for capacitors as new measurements become available. This capability increases the technology readiness level of prognostics applied to electrolytic capacitors.

7.1. Ongoing and future work

The results presented here are based on accelerated life test data and on the accelerated life timescale. Further research will focus on development of functional mappings that will translate the accelerated life timescale into real usage conditions time-scale, where the degradation process dynamics will be slower, and subject to several types of stresses.

The performance of the proposed exponential-based degradation model is satisfactory for this study based on the quality of the model fit to the experimental data and the RUL prediction performance as compared to ground truth. As part of future work we will also focus on the exploration of additional models based on the physics of the degradation process and larger sample size for aged devices.

Additional experiments are currently underway to increase the number of test samples. This will greatly enhance the quality of the model, and guide the exploration of additional degradation-models, where the loading conditions and the environmental conditions are also accounted for towards degradation dynamics.

ACKNOWLEDGMENT

This work was funded by the NASA Aviation Safety Program, SSAT project.

NOMENCLATURE

C_I	Ideal capacitance value for an ideal capacitor
C_R	Real capacitor value for a non-ideal capacitor model
R_E	Equivalent series resistance of the capacitor
$C_l(k)$	Capacitance percentage loss at time t_k (state variable)
T_i	Validation test on capacitor i
$\mathcal{M}_i$	Nominal model for a component or system
$\mathcal{D}_i$	Degradation model for a capacitor
R_L	Load resistance on electrical overstress system
V_L	Load voltage on electrical overstress system
V_o	Electrical overstress voltage in aging system
Z_I	Ideal capacitor impedance
Z	Capacitor impedance for non-ideal capacitor model $\mathcal{M}_1$

REFERENCES

Bhatti, U., & Ochieng, W. (2007). Failure modes and models for integrated GPS/INS systems. *The Journal of Navigation*, 60, 327.

Celaya, J., Kulkarni, C., Biswas, G., & Goebel, K. (2011a, September). A Model-based Prognostics Methodology for Electrolytic Capacitors Based on Electrical Overstress Accelerated Aging. In *Proceedings of the Annual Conference of the Prognostics and Health Management Society*. Montreal, Canada.

Celaya, J., Kulkarni, C., Biswas, G., & Goebel, K. (2011b, March). Towards Prognostics of Electrolytic Capacitors. In *In Proceedings of the AIAA Infotech@Aerospace Conference*. St. Louis, MO.

Celaya, J., Kulkarni, C., Biswas, G., & Goebel, K. (2012,

January). Prognostic and Experimental Techniques for Electrolytic Capacitor Health Monitoring. In *Proceedings of the The Annual Reliability and Maintainability Symposium (RAMS)*. Reno, Nevada.

Chatfield, C. (2003). *The Analysis of Time Series: An Introduction* (6th ed.). Chapman and Hall/CRC.

Goodman, D., Hofmeister, J., & Judkins, J. (2007). Electronic prognostics for switched mode power supplies. *Microelectronics Reliability*, *47*(12), 1902-1906.

Gu, J., Azarian, M. H., & Pecht, M. G. (2008). Failure prognostics of multilayer ceramic capacitors in temperature-humidity-bias conditions. In *International Conference on Prognostics and Health Management* (p. 1-7). Denver, CO.

International Standard IEC 60384-4-1: Fixed capacitors for use in electronic equipment (3rd ed.). (2007). International Electrotechnical Commission (IEC).

Judkins, J. B., Hofmeister, J., & Vohnout, S. (2007). A Prognostic Sensor for Voltage Regulated Switch-Mode Power Supplies. In *IEEE Aerospace Conference* (p. 1-8).

Kulkarni, C., Biswas, G., Bharadwaj, R., & Kim, K. (2010). Effects of Degradation in DC-DC Converters on Avionics Systems: A Model Based Approach. In *Machinery Failure Prevention Technology Conference, MFPT*.

Kulkarni, C., Biswas, G., & Koutsoukos, X. (2009). A prognosis case study for electrolytic capacitor degradation in DC-DC converters. In *Proceedings of the Annual Conference of the Prognostics and Health Management Soceity*. San Diego, CA.

Kulkarni, C., Biswas, G., Koutsoukos, X., Celaya, J., & Goebel, K. (2010). Integrated diagnostic/prognostic experimental setup for capacitor degradation and health monitoring. In *IEEE AUTOTESTCON*.

Kulkarni, C., Biswas, G., Koutsoukos, X., Goebel, K., & Celaya, J. (2010a). Experimental Studies of Ageing in Electrolytic Capacitors. In *Proceedings of the Annual Conference of the Prognostics and Health Management Soceity*.

Kulkarni, C., Biswas, G., Koutsoukos, X., Goebel, K., & Celaya, J. (2010b). Physics of Failure Models for Capacitor Degradation in DC-DC Converters. In *The Maintenance and Reliability Conference, MARCON*.

Kulkarni, C., Celaya, J., Goebel, K., & Biswas, G. (2012a, September). Bayesian Framework Approach for Prognostic Studies in Electrolytic Capacitor under Thermal Overstress Conditions. In *Proceedings of Annual Conference of the Prognostics and Health Management Society*. Minneapolis, MN.

Kulkarni, C., Celaya, J., Goebel, K., & Biswas, G. (2012b). Physics Based Electrolytic Capacitor Degradation Models for Prognostic Studies under Thermal Overstress. In *Proceedings of First European Conference of the Prognostics and Health Management Soci-*

ety. Dresden, Germany.

Kulkarni, C., Celaya, J., Goebel, K., & Biswas, G. (2012c, June). Prognostics Health Management and Physics based failure Models for Electrolytic Capacitors. In *Proceedings of American Institute of Aeronautics and Astronautics, AIAA Infotech@Aerospace Conference*. Garden Grove, CA.

MIL-C-62F. (2008). *General specification for capacitors, fixed, electrolytic (dc. aluminum, dry electrolyte, polarized),*. Military Specification. Department of Defense.

Nie, L., Azarian, M. H., Keimasi, M., & Pecht, M. (2007). Prognostics of ceramic capacitor temperature-humidity-bias reliability using Mahalanobis distance analysis. *Circuit World*, *33*(3), 21 - 28.

Orsagh, R., Brown, D., Roemer, M., Dabnev, T., & Hess, A. (2005). Prognostic health management for avionics system power supplies. In *IEEE Aerospace Conference* (p. 3585-3591).

Saha, B., Goebel, K., & Christophersen, J. (2009). Comparison of Prognostic Algorithms for Estimating Remaining Useful Life of Batteries. *IEEE Transactions of the Institute of Measurement and Control*, *31*(3-4), 293-308.

Saxena, A., Celaya, J., Saha, B., Saha, S., & Goebel, K. (2010). Metrics for offline evaluation of prognostic performance. *International Journal of Prognostics and Health Management*, *1-1*(1).

Stengel, R. F. (1994). *Optimal Control and Estimation*. Dover Books on Advanced Mathematics.

Dr. José R. Celaya is a research scientist with SGT Inc. at the Prognostics Center of Excellence, NASA Ames Research Center. He received a Ph.D. degree in Decision Sciences and Engineering Systems in 2008, a M. E. degree in Operations Research and Statistics in 2008, a M. S. degree in Electrical Engineering in 2003, all from Rensselaer Polytechnic Institute, Troy New York; and a B. S. in Cybernetics Engineering in 2001 from CETYS University, México.

Chetan S. Kulkarni is a Research Assistant at ISIS, Vanderbilt University. He received the M.S. degree in EECS from Vanderbilt University, Nashville, TN, in 2009, where he is currently a Ph.D. candidate and received a B. E. degree in Electronics and Electrical Engineering from University of Pune, India in 2002.

Dr. Kai Goebel is a deputy area lead of the Discovery and Systems Health Technology Area at NASA Ames Research Center. He also coordinates the Prognostics Center of Excellence and is the Technical Lead for Prognostics and Decision Making in NASAs System-wide Safety and Assurance Technologies Project. Prior to joining NASA in 2006, he was a senior research scientist at General Electric Corporate Research and Development center since 1997. Dr. Goebel received his Ph.D. at the University of California at Berkeley in 1996. He has carried out applied research in the areas of real time monitoring, diagnostics, and prognostics and he has fielded numerous applications for aircraft engines, transportation systems, medical systems, and manufacturing systems. He holds 17 patents and has co-authored more than 250 technical papers in the field of IVHM. Dr. Goebel was an adjunct professor of the CS Department at Rensselaer Polytechnic Institute (RPI), Troy, NY, between 1998 and 2005 where he taught classes in Soft Computing and Applied Intelligent Reasoning Systems. He has been the co-advisor of 6 Ph.D. students. Dr. Goebel is a member of several professional societies, including ASME, AAAI, AIAA, IEEE, VDI, SAE, and ISO. He was the General Chair of the Annual Conference of the PHM Society, 2009, has given numerous invited and keynote talks and held many chair positions at the PHM conference and the AAAI Annual meetings series. He is currently member of the board of directors of the PHM Society and associate editor of the International Journal of PHM.

Gautam Biswas is a Professor of Computer Science, Computer Engineering, and Engineering Management in the EECS Department and a Senior Research Scientist at the Institute for Software Integrated Systems (ISIS) at Vanderbilt University. He has an undergraduate degree in Electrical Engineering from the Indian Institute of Technology (IIT) in Mumbai, India, and M.S. and Ph.D. degrees in Computer Science from Michigan State University in E. Lansing, MI.

Prof. Biswas conducts research in Intelligent Systems with primary interests in hybrid modeling, simulation, and analysis of complex embedded systems, and their applications to diagnosis, prognosis, and fault-adaptive control. As part of this work, he has worked on fault diagnosis and fault-adaptive control of secondary sodium cooling systems for nuclear reactors, automobile engine coolant systems, fuel transfer and avionics systems for aircraft, Advanced Life Support systems and power distribution systems for NASA. He has also initiated new projects in health management of complex systems, which includes online algorithms for distributed monitoring, diagnosis, and prognosis. More recently, he is working on data mining for diagnosis, and developing methods that combine model-based and data-driven approaches for diagnostic

and prognostic reasoning. This work, in conjunction with Honeywell Technical Center and NASA Ames, includes developing sophisticated data mining algorithms for extracting causal relations amongst variables and parameters in a system. In other research projects, he is involved in developing simulation-based environments for learning and instruction. His industrial collaborators include Airbus, Honeywell Technical Center, and Boeing Research and Development. He has published extensively, and has over 300 refereed publications.

Dr. Biswas is an associate editor of the IEEE Transactions on Systems, Man, and Cybernetics, Prognostics and Health Management, and Educational Technology and Society journal. He has served on the Program Committee of a number of conferences, and most recently was Program co-chair for the 18th International Workshop on Principles of Diagnosis and Program Coordination Chair for the 20th International Conference in Computers in Education. He is currently serving on the Executive committee of the Asia Pacific Society for Computers in Education and is the IEEE Computer Society representative to the Transactions on Learning Technologies steering committee. He is also serving as the Secretary/Treasurer for ACM Sigart. He is a senior member of the IEEE Computer Society, ACM, AAAI, and the Sigma Xi Research Society.

A. PROGNOSTICS VALIDATION RESULTS

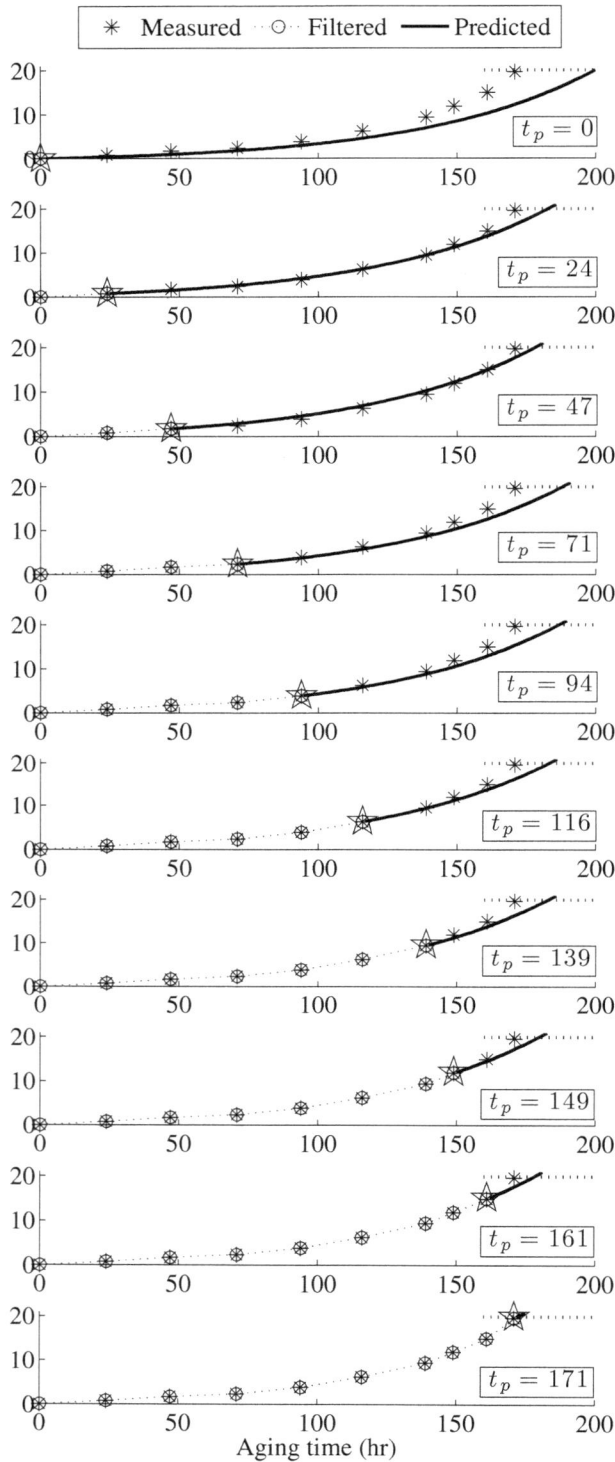

Figure 17. T_2: Health state estimation and forecasting of capacitance loss (%) at different times t_p during the aging time; $t_p = [0, 24, 47, 71, 94, 116, 139, 149, 161, 171]$.

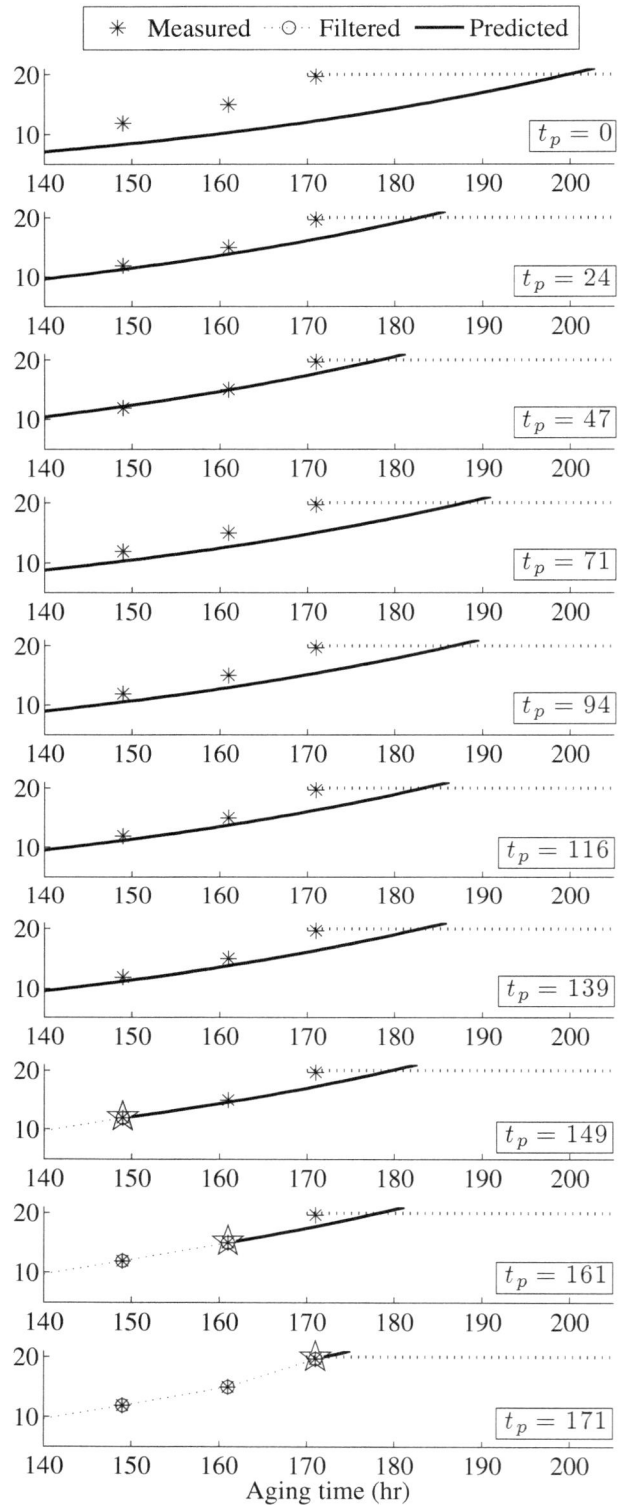

Figure 18. T_2: Detail of the health state estimation and forecasting of capacitance loss (%) at different times t_p during the aging time; $t_p = [0, 24, 47, 71, 94, 116, 139, 149, 161, 171]$.

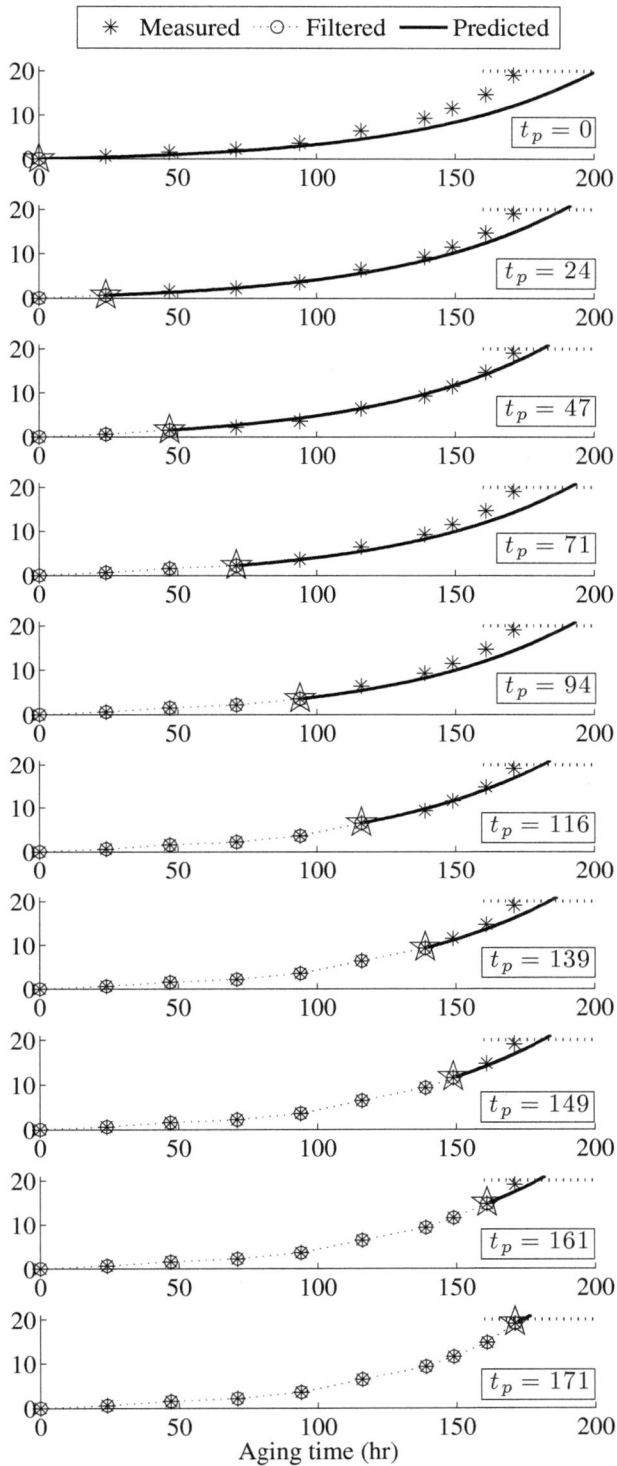

Figure 19. T_3: Health state estimation and forecasting of capacitance loss (%) at different times t_p during the aging time; $t_p = [0, 24, 47, 71, 94, 116, 139, 149, 161, 171]$.

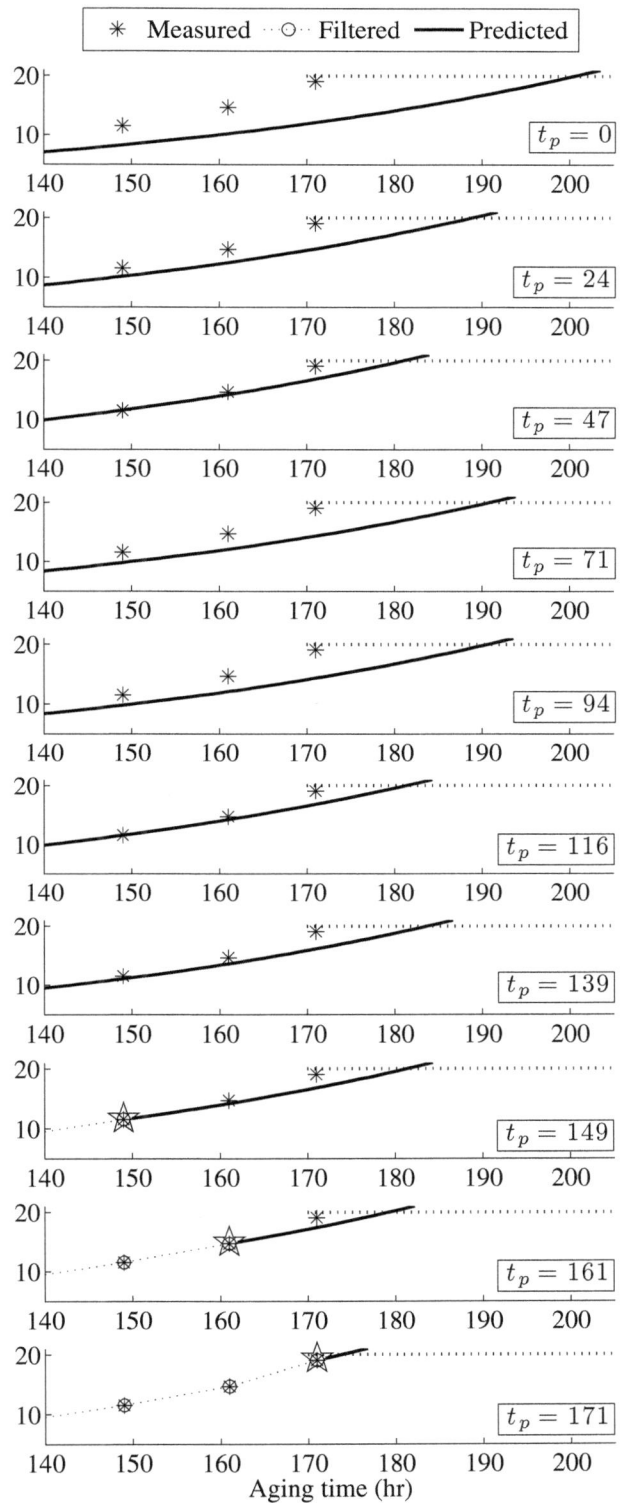

Figure 20. T_3: Detail of the health state estimation and forecasting of capacitance loss (%) at different times t_p during the aging time; $t_p = [0, 24, 47, 71, 94, 116, 139, 149, 161, 171]$.

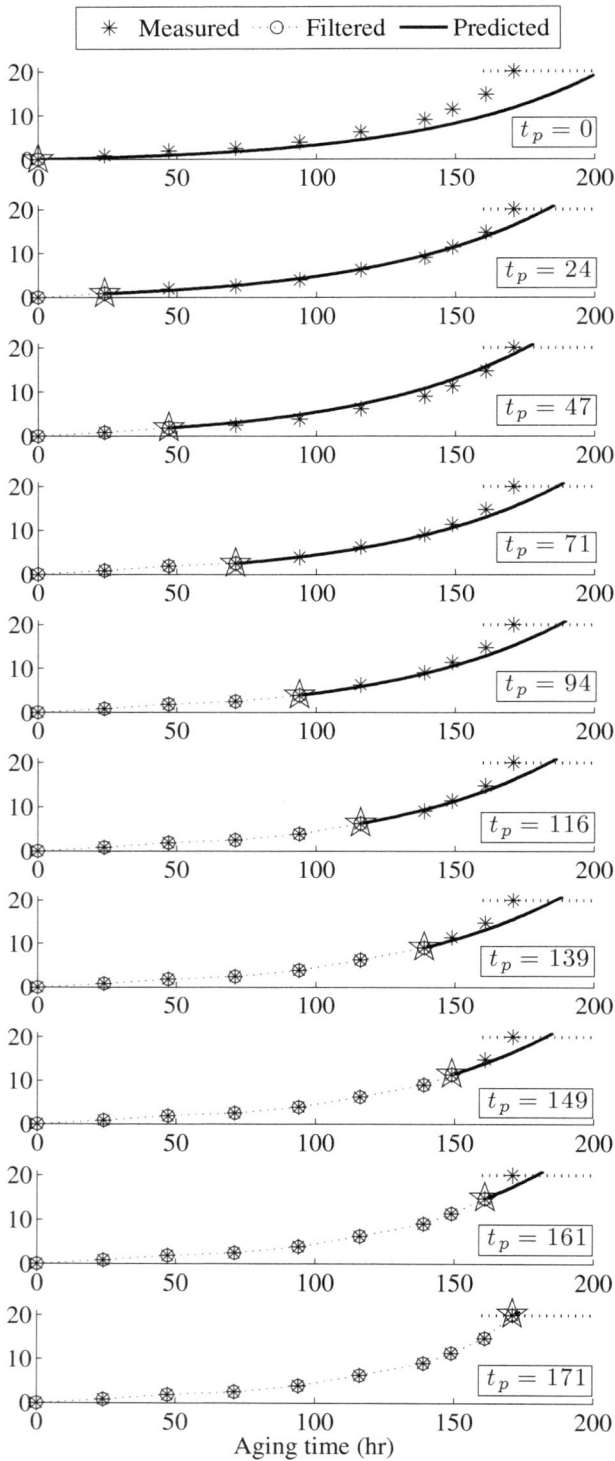

Figure 21. T_4: Health state estimation and forecasting of capacitance loss (%) at different times t_p during the aging time; $t_p = [0, 24, 47, 71, 94, 116, 139, 149, 161, 171]$.

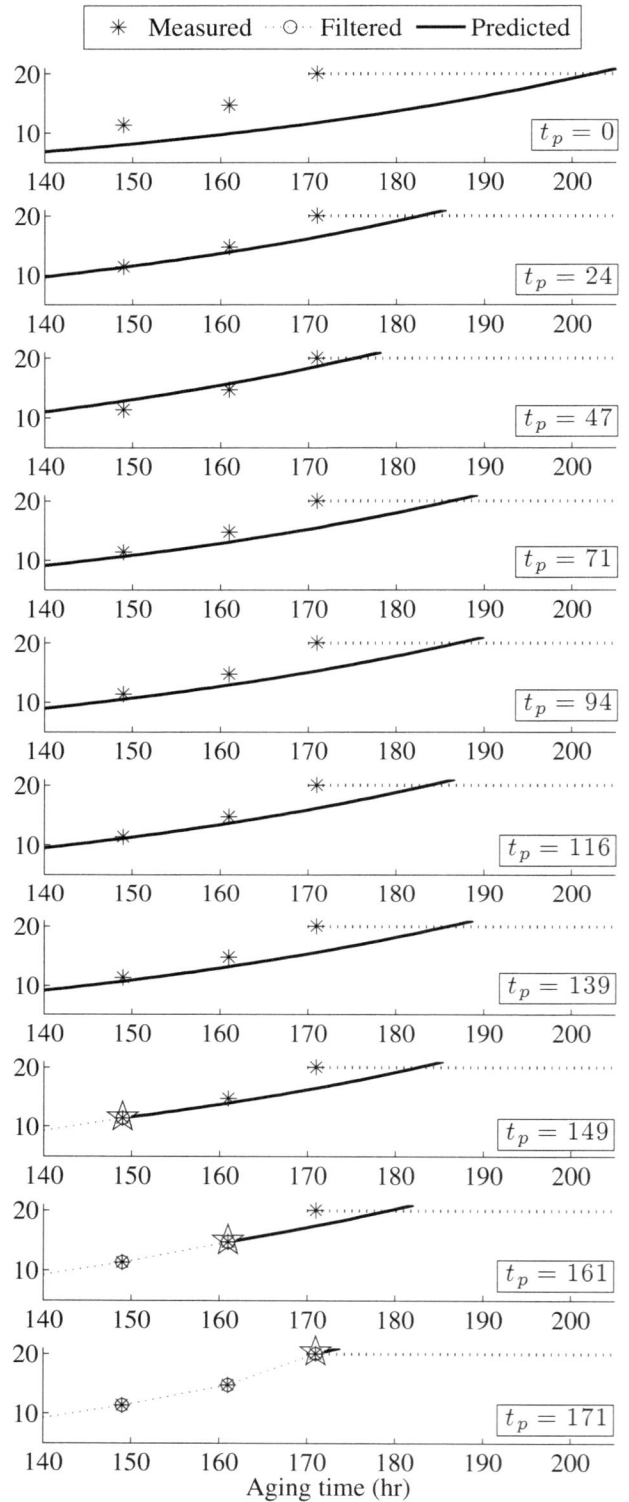

Figure 22. T_4: Detail of the health state estimation and forecasting of capacitance loss (%) at different times t_p during the aging time; $t_p = [0, 24, 47, 71, 94, 116, 139, 149, 161, 171]$.

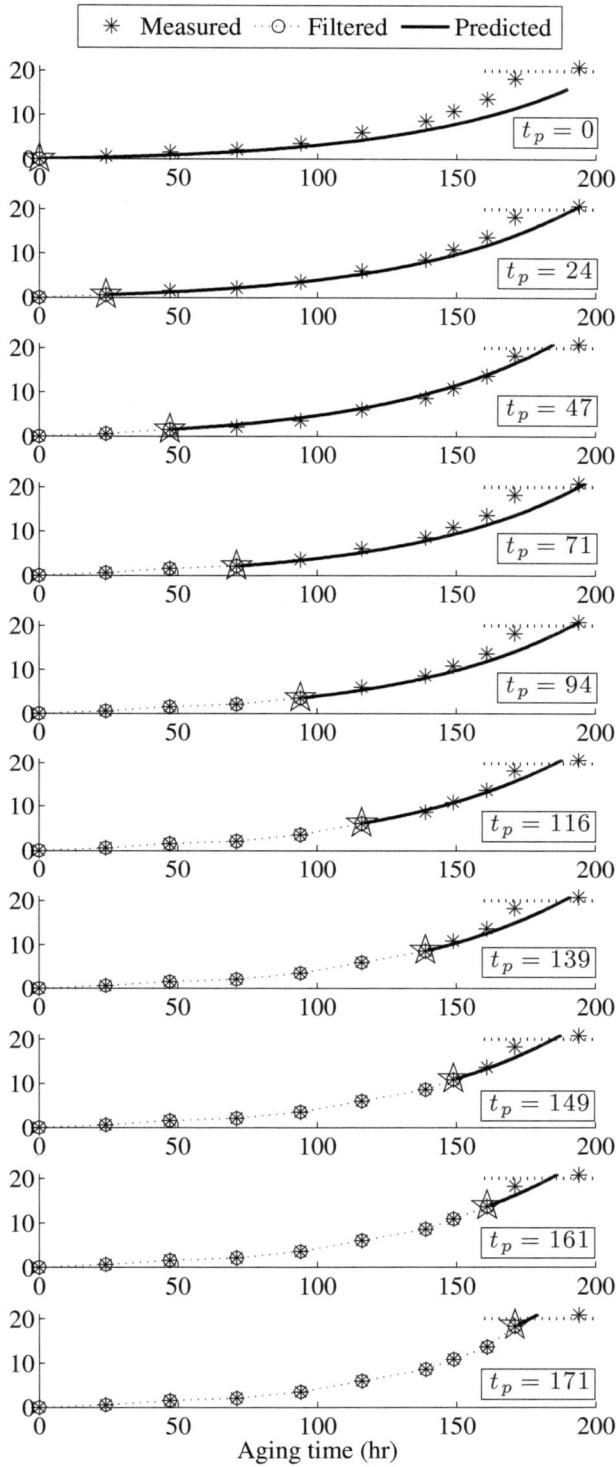

Figure 23. T_5: Health state estimation and forecasting of capacitance loss (%) at different times t_p during the aging time; $t_p = [0, 24, 47, 71, 94, 116, 139, 149, 161, 171]$.

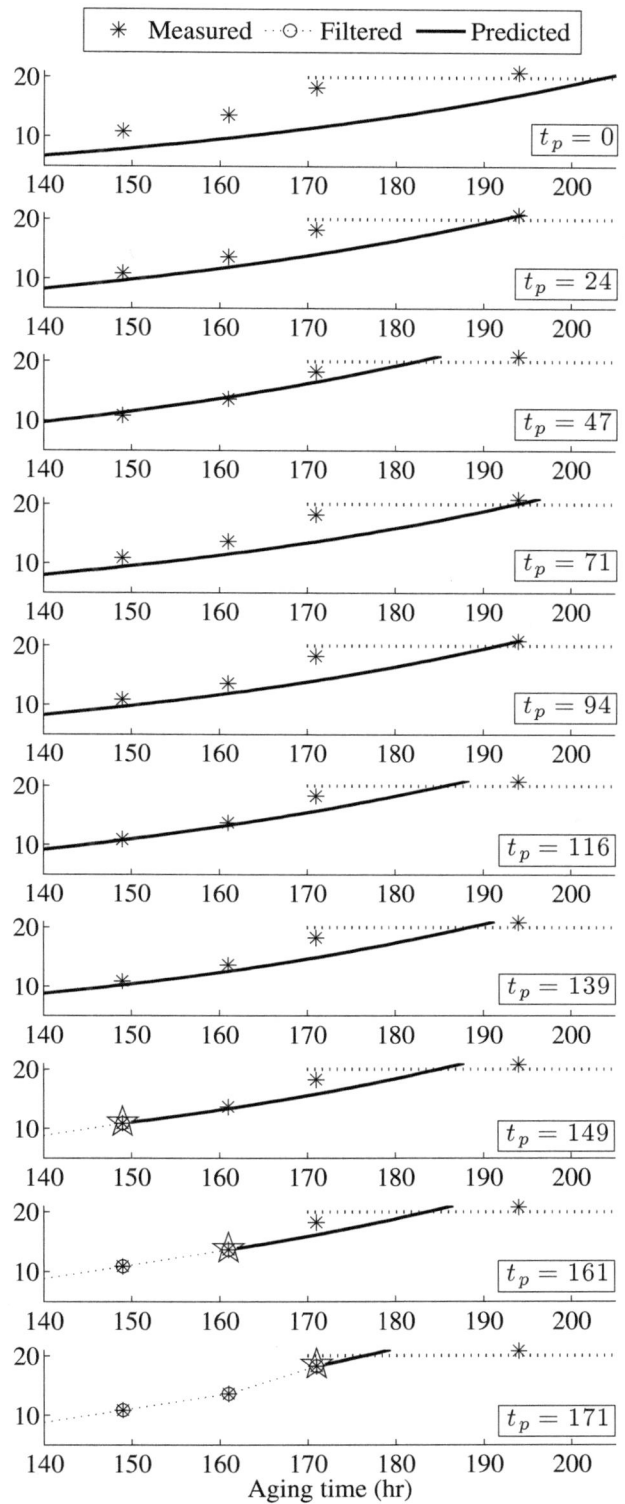

Figure 24. T_5: Detail of the health state estimation and forecasting of capacitance loss (%) at different times t_p during the aging time; $t_p = [0, 24, 47, 71, 94, 116, 139, 149, 161, 171]$.

A. PROGNOSTICS ALPHA-LAMBDA PERFORMANCE METRIC

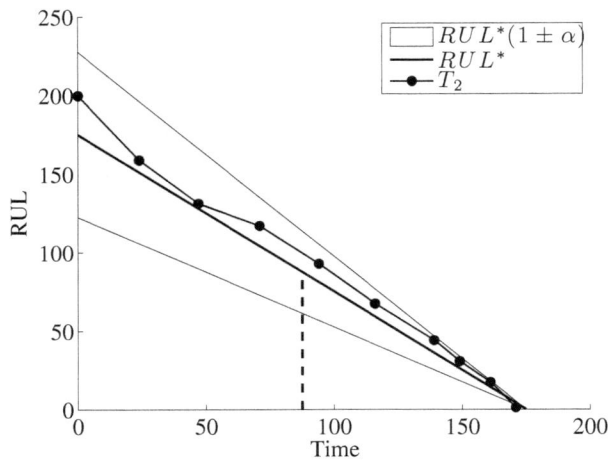

Figure 25. T_2: Alpha-Lambda Prognostics Metric ($\lambda = 0.5$ and $\alpha = 0.3$).

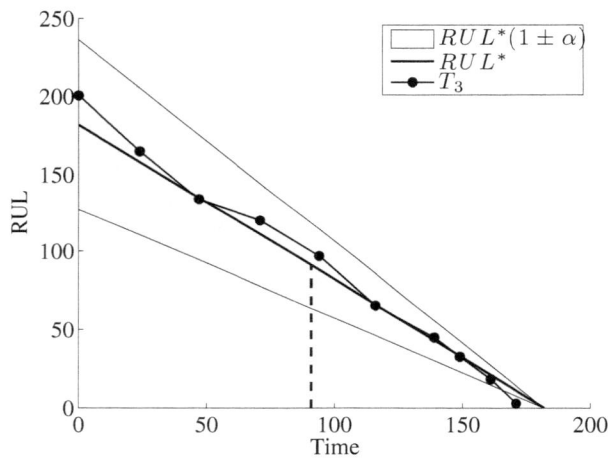

Figure 26. T_3: Alpha-Lambda Prognostics Metric ($\lambda = 0.5$ and $\alpha = 0.3$).

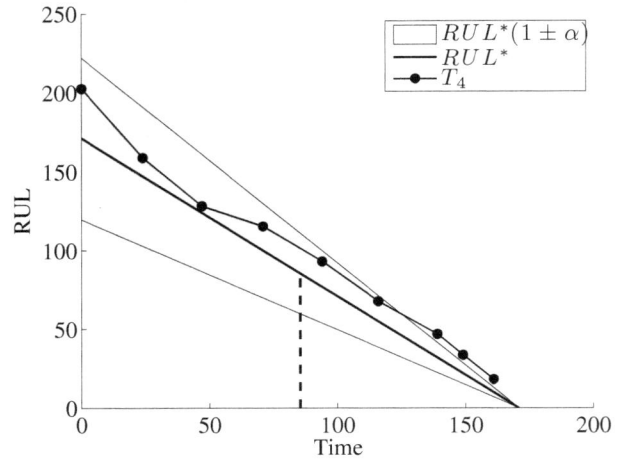

Figure 27. T_5: Alpha-Lambda Prognostics Metric ($\lambda = 0.5$ and $\alpha = 0.3$).

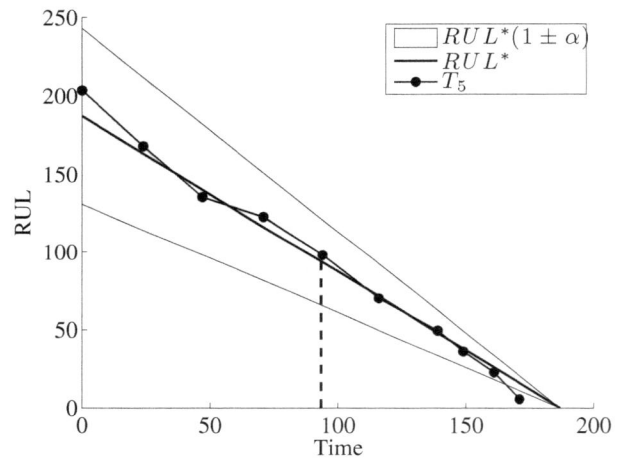

Figure 28. T_5: Alpha-Lambda Prognostics Metric ($\lambda = 0.5$ and $\alpha = 0.3$).

INTERNATIONAL JOURNAL OF PROGNOSTICS AND HEALTH MANAGEMENT, VOL.3 (2012)

Diagnostics of Local Tooth Damage in Gears by the Wavelet Technology

L. Gelman[1], I. Petrunin[2], I. K. Jennions[3], and M. Walters[4]

[1,2] *School of Engineering, Cranfield University, Cranfield, Bedfordshire, MK43 0AL, UK*
l.gelman@cranfield.ac.uk
i.petrunin@cranfield.ac.uk

[3] *IVHM Centre, Cranfield University, Cranfield, Bedfordshire, MK43 0AL, UK*
i.jennions@cranfield.ac.uk

[4] *Rolls-Royce, Derby, DE24 8BJ*
mark.walters@rolls-royce.com

ABSTRACT

The chipped gear tooth appearance is normally a result of the initial fatigue damage in a tooth. It is a special gear failure mode and differs from local fatigue damage of gear teeth. Therefore, diagnosis of chipped gear tooth requires a special investigation. Recently, the novel gear damage diagnosis technology, based on the wavelet transform was proposed and successfully applied for diagnosis of the early stage fatigue damage. The proposed technology is applied in this study for diagnostics of a partly-missing (chipped) tooth in a gear of the Machine Fault Simulator (MFS). The advanced automatic technology for the time synchronous averaging of the raw gear vibrations has been employed in this study; this technology does not require speed data. An advanced decision making technique based on use of the likelihood ratio allowed for the continuous correct diagnosis of chipped teeth throughout the recorded data without false alarms and missed detections. The likelihood ratio was obtained using the Gaussian models of the data for classes "undamaged" and "damaged".

1. INTRODUCTION

The vibration-based gear diagnostics has been the most popular monitoring technique because of its high effectiveness. Multiple methods for vibration diagnostics of local fatigue damage in gears (i.e. pitting, etc.) are proposed and investigated including statistical analysis of raw vibration signal analysis techniques (Maynard, 1999, Lebold, McClintic, Campbell, Byington, & Maynard, 2000, Lei, Zuo, He, & Zi, 2010), spectral analysis (Yesilyurt, 2003), demodulation methods (McFadden, 1986, Gelman et al., 2005, Combet, Gelman, Anuzis, & Slater, 2009), residual analysis (Combet at al., 2009, Wang, Ismail, & Golnaraghi, 2001), adaptive filtering (Brie, Tomczak, Oehlmann, & Richard, 1997, Lee & White, 1998, Combet & Gelman, 2009), use of AR model (Wang & Wong, 2002), inverse filtering (Lee & Nandi, 2000, Endo & Randall, 2007), time frequency (TF) analysis (Wang & McFadden, 1993, Forrester, 1996, Choy, Polyshchuk, Zakrajsek, Handschuh, & Townsend, 1996, Wang & McFadden, 1996, Loutridis, 2006) and time-scale analysis (Wang et al., 2001, Halima, Shoukat Choudhuryb, Shaha, & Zuoc, 2008, Dalpiaz, Rivola, & Rubini, 2000, Lin & Zuo, 2003, Combet, Gelman, & LaPayne, 2012).

The majority of the these methods are based on the residual signal as classically obtained after the removal of the mesh harmonic components from the gear vibration signal processed using the time synchronous average (TSA) (Stewart, 1977, McFadden, 1987).

Although the chipped gear tooth (i.e. partly missing tooth) normally appears as a result of the initial fatigue damage in a tooth, it is a special gear failure mode and differs from local fatigue damage of gear teeth. Therefore, normally, diagnosis of chipped gear tooth requires a special investigation.

The literature search revealed a limited number of publications in which the detection of the chipped tooth was investigated using advanced signal processing techniques, such as joint amplitude and frequency demodulation analysis (Feng, Liang, Zhang, & Hou, 2012), bispectral binary images (Jiang, Liu, Li & Tang, 2011), and empirical mode decomposition (Zamanian & Ohadi, 2011).It is believed that this topic should be further investigated by evaluation of suitability of novel damage diagnosis techniques for diagnosis of the chipped gear teeth.

Recently, the novel gear damage diagnosis technology, based on the wavelet transform, was proposed and successfully applied for diagnosis of the fatigue damage, micro-pitting (i.e. 0.3-0.7% relative pitting size) in gear teeth (Gryllias, Gelman, Shaw & Vaidhianathasamy, 2010). However, the full capabilities of this technology are not yet investigated. While being effective for early diagnosis of fatigue damage in teeth, the technology was not investigated for diagnosis of the chipped gear teeth.

Therefore, the main aim of the present paper is to investigate diagnostics of partly missing (chipped) teeth using the novel wavelet technology. The novelty of the paper is investigation for the first time of diagnosis of the chipped tooth using the novel wavelet technology.

2. NOVEL DIAGNOSTIC FEATURE BASED ON THE WAVELET TRANSFORM

The wavelet transform (WT) can be presented in the form (Yan, Miyamoto & Brühwiler, 2006):

$$W(a,b) = \frac{1}{\sqrt{|a|}} \int_{-\infty}^{+\infty} x(t) \psi^* \left(\frac{t-b}{a} \right) dt, \quad (1)$$

where $x(t)$ is a time domain signal. ψ is the mother wavelet function, a is the scale variable, b is the time shift, * denotes the complex conjugation.

In this study, the complex Morlet wavelet function is used:

$$\psi(t) = \frac{1}{\sqrt{\pi f_b}} \left(e^{j2\pi f_c t} - e^{-f_b(\pi f_c)^2} \right) e^{-t^2/f_b},$$

where f_b is the bandwidth parameter, f_c is the central frequency, j is the imaginary unit.

It is known that local gear faults produce impacts; frequencies of these impacts are localized in a specific band. The normalized wavelet transform integrated in this band is proposed (Gryllias et al., 2010) as the instantaneous diagnostic feature.

$$y = \frac{\int_B |W(a,b)|^2 df}{B},$$

where $|W(a,b)|$ denotes the magnitude of the wavelet transform; $B = [f_{min}, f_{max}]$ is the frequency band for the integration; f_{min} is the minimum frequency of the band B; f_{max} is the maximum frequency of the band B.

The choice of the integration bandwidth B needs to be adapted according to the frequencies of impact produced by a fault. The average feature, the mean of the instantaneous feature in the mesh period, is used in this study.

The bandwidth parameter f_b can be presented as follows:

$$f_b = \left(\frac{t_B f_c}{2} \right)^2,$$

The dimensionless product $t_B f_c$ corresponds to the number of oscillations of the Morlet wavelet within its half power time-width estimated at the -3 dB level. This product defines the shape of the wavelet, and thus the balance between the time and frequency resolutions of the wavelet transform.

In order to maximize the effectiveness of diagnostics of local tooth fault, the time resolution of the wavelet analysis should match with the duration of the fault-induced impact. Normally, the impact duration is of the order of the mesh period. Therefore, the dimensionless window length parameter $t_B f_c$ was set so in order that the time-width at_B of the analyzing wavelet at scale a matches with the meshing period T_m of the gear for all scales. This implies that $at_B = t_B f_c / f \approx T_m$ for all frequencies within the bandwidth $B = [f_{min}, f_{max}]$. By considering the middle frequency only, this condition may be approximated as:

$$t_B f_c \approx T_m \frac{1}{2} (f_{min} + f_{max}).$$

3. THE NOVEL WAVELET TECHNOLOGY FOR THE DIAGNOSTICS OF THE GEAR LOCAL DAMAGE

A schematic of the damage diagnosis technology, based on the novel residual signal use is shown in Fig. 1.

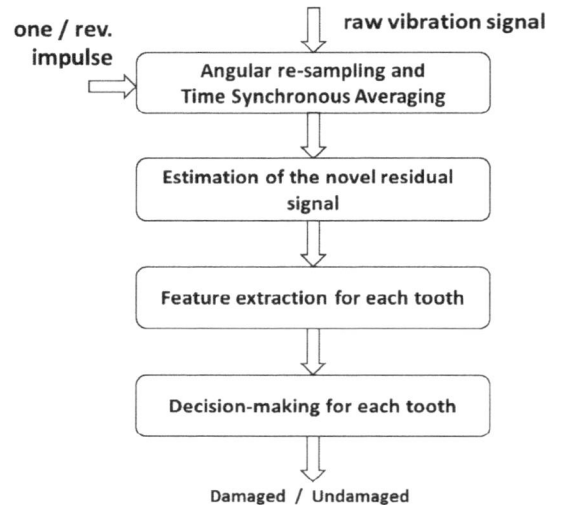

Figure 1. Schematic of the damage diagnosis technology.

The first stage of the technology used in this paper is the angular re-sampling of the raw vibration signal using the estimate of the shaft speed. The speed estimate can be obtained using the one per rev signal from the tachometer;

however, in some cases the speed can be extracted from the vibration data without need of the tachometer signal. In the present paper, the advanced automatic technology for the time synchronous averaging of the raw gear vibrations (Combet & Gelman, 2007) has been employed; this technology does not require speed data. The angular re-sampling is performed automatically using the mesh components of the vibration data (Combet & Gelman, 2007).

The TSA is then applied to the re-sampled signal. In order to estimate the signal duration required for the TSA, the dependency of the ratio between the averaged variance of the TSA signal and the variance of the re-sampled vibration signal versus the number of averages (Combet & Gelman, 2007) was estimated. The length of the raw signal required for the TSA was estimated as the number of averages at which the ratio of variances begins to demonstrate relatively low attenuation (Fig. 2).

The novel residual signal is obtained from the TSA signal by removing not only mesh harmonics but also low order shaft harmonics. This approach first introduced in (Combet et al., 2009) improves diagnostics effectiveness as low order shaft harmonics normally have relatively high amplitudes and, therefore, could mask impacts created by local gear faults.

The decision making procedure is based on the likelihood ratio estimated from training data with a priori known classification. During the testing, values of the likelihood ratio are accumulated and compared to the threshold in order to make the final decision.

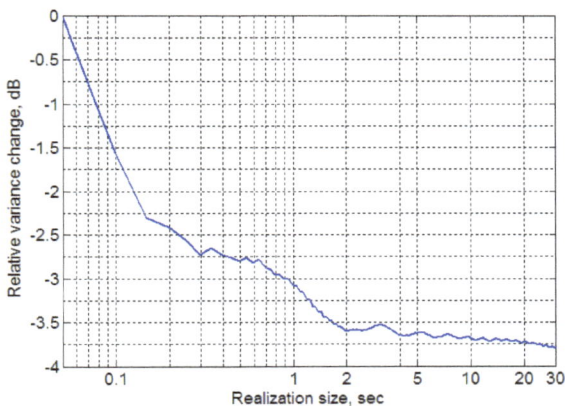

Figure 2. Estimation of the relative variance change of the TSA signal vs realization size.

4. THE EXPERIMENT DESCRIPTION

The experimental set up is based on the MFS test rig (Fig. 3) equipped with a one-stage gearbox. Two pinions were used for the test: a pinion with no damage on teeth and pinion with a chipped tooth (Fig. 4). The test was performed at a constant motor shaft speed of 3000 rpm-which

corresponds to 1200 rpm on the tested gear shaft. The load on the gear was applied by a magnetic brake system with the torque value of 0.68 N*m.

Vibrations from three channels for axial, horizontal and vertical directions were recorded using 3 Endevco 7251A-100 accelerometers installed on the gearbox case. Signal conditioning was performed using an Endevco 133 signal conditioner and analogue active anti-aliasing filters Kemo PocketMaster 1600. The cut-off frequency of filters was set at 10 kHz. The speed estimation was obtained using a laser speed sensor and reflective tape, attached to the driven wheel of the belt transmission (Fig. 3). All signals were recorded using National Instruments data acquisition card NI DAQ-6062E at 25 kHz sampling frequency.

The whole data set is represented by 7 records of approximately 3 minutes duration each with 10 minute intervals between them for damaged and undamaged gears.

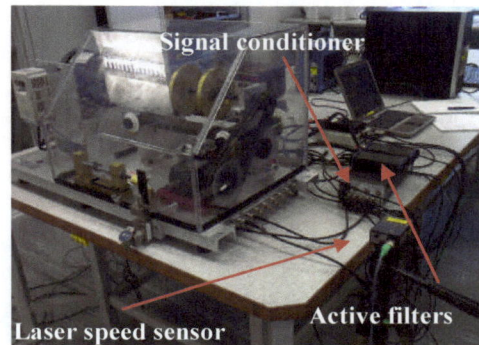

Figure 3. The test rig (MFS).

Figure 4. The gear with partly missing tooth.

The length of realizations for the residual signal estimation using the TSA was selected as 20s according to Fig. 2. Therefore, the whole data set of 7 records contains 64 realizations. For 20s realization, 1200 rpm shaft speed, 18 teeth and 25 kHz sampling frequency each tooth is represented by approximately 70 samples of the vibration data. While this number looks reasonably large for the considered diagnostics task it might be important to consider the influence of this number on diagnostics efficiency in the future.

5. DATA PROCESSING AND DIAGNOSTICS

The wavelet transform of the novel residual signal was computed for a number of realizations in order to evaluate the frequency bandwidth of impacts produced by the fault. The window length parameter $t_B f_c$ that matches the mesh period at the middle frequency of the bandwidth was selected.

Wavelet transform Eq. (1) of the novel residual signal for bandwidth 500 Hz-3 kHz averaged over nine realizations for data from vertical channel are shown in Fig. 5.

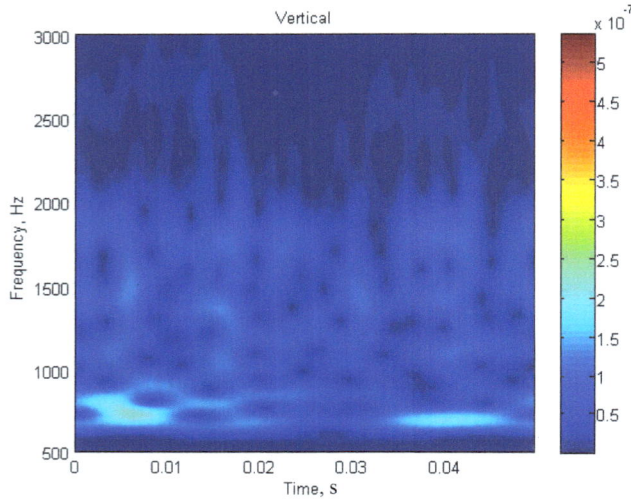

(a) Undamaged

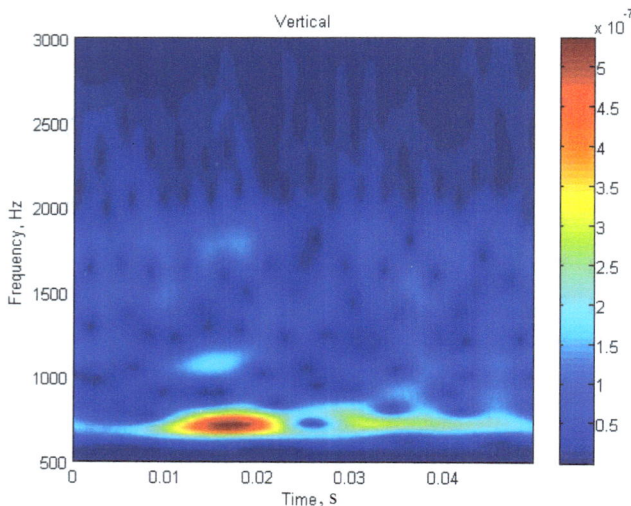

(b) Damaged

Figure 5. The averaged wavelet transforms of the novel residual signal for undamaged (a) and damaged (b) gears.

It can be seen from Fig. 5 the increased wavelet coefficients for damaged gear at frequency around 700 Hz and time 0.011-0.023s, which generated by the chipped tooth. A channel selection for diagnostics was performed based on the separation of diagnostic features between data from undamaged and damaged gears. The separation was estimated as the maximum value of the Fisher criterion (Webb, 1999) over the rotation period using all available data (i.e. 64 realizations for undamaged gear and 64 realizations for damaged gear). The Fisher criterion values were estimated from the expression as follows:

$$F = \left(m_1 - m_2\right)^2 \big/ \left(\sigma_1^2 + \sigma_2^2\right),$$

where m_1, m_2 and σ_1^2, σ_2^2 are mean values and variances of diagnostic features from each realization for damaged and undamaged gears, respectively.

The same methodology was used for the wavelet bandwidth selection. Results of the analysis are shown in Table 1.

	Bandwidth 1: 10 Hz-10 kHz	Bandwidth 2: 500 Hz-3 kHz	Bandwidth 3: 600 Hz-850 Hz
Axial	2.7	7.8	20.3
Horizontal	2.2	5.8	14.4
Vertical	3.3	7.3	37.2

Table 1. Values of the Fisher criterion.

As the result of the analysis, the vertical direction of accelerometer and the narrowest bandwidth (600 Hz – 850 Hz) were selected for diagnostics.

In addition to removing the gear mesh harmonics, the first eight shaft orders were removed from the TSA for estimation of the residual signals.

Plots of the diagnostic features for each tooth of the pinion using the data from the vertical channel for undamaged and damaged gears are shown in Fig 6 (a) and (b) respectively. One can see that for the damaged gear tooth 7 (i.e. the chipped tooth) has higher feature values.

(a) Undamaged

(b) Damaged

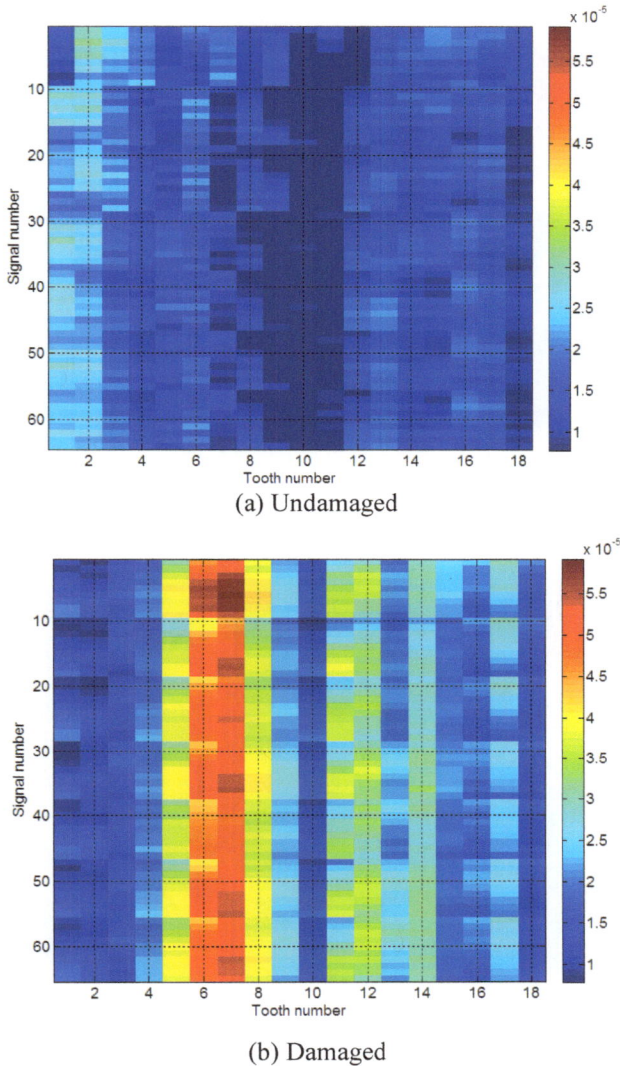

Figure 6. The diagnostic features vs. tooth number and realization number.

The training data for the damage diagnosis were selected as follows:

- Class "damaged" was presented by 32 diagnostic features of damaged tooth 7 in 32 gear vibration signals, i.e. every other realization of the total amount of 64 realizations.

- Class "undamaged" was presented by diagnostic features of all 18 teeth of 32 realizations for the undamaged gear and diagnostic features of remaining 17 teeth (i.e. except the tooth number 7) of 32 realizations for the damaged gear; in total, 1120 diagnostic features were used for the "undamaged class".

Observing the uni-modal shape of diagnostic features distributions, the likelihood ratio was obtained using the

Gaussian models of the data for classes "undamaged" and "damaged". To build the models, the corresponding values of mean and variance were estimated for each class. Models of the probability density functions of the diagnostic features and the logarithm of the likelihood ratio are shown in Fig. 7.

The likelihood ratio is estimated using the selected damaged and undamaged training data. The testing data are processed using the accumulated likelihood ratio given by:

$$\Lambda_N = \log \frac{p(x_1,...,x_N \mid \omega_{dam})}{p(x_1,...,x_N \mid \omega_{undam})} =$$

$$= \log \frac{\prod_i p(x_i \mid \omega_{dam})}{\prod_i p(x_i \mid \omega_{undam})} = \sum_{i=1}^{N_{acc}} \log \frac{p(x_i \mid \omega_{dam})}{p(x_i \mid \omega_{undam})},$$

where $p(x_i \mid \omega_{dam})$ and $p(x_i \mid \omega_{undam})$ are the conditional probability density functions of the diagnostic features for damaged ω_{dam} and undamaged ω_{undam} classes; i=1,..., N_{acc}, N_{acc} is the number of accumulations.

The decision making rule for damage diagnosis using the sequence of realizations is as follows:

$$\Lambda_N \geq thr_b,$$

where thr_b is the threshold for the accumulated likelihood ratio.

The test was performed using the test data represented by another half (32 realizations) of the whole data set (64 realizations). The values of $thr_b = 23$ and $N_{acc} = 5$ were obtained by optimization of the damage diagnosis procedure using the minimum of the total error probability as the optimization criterion.

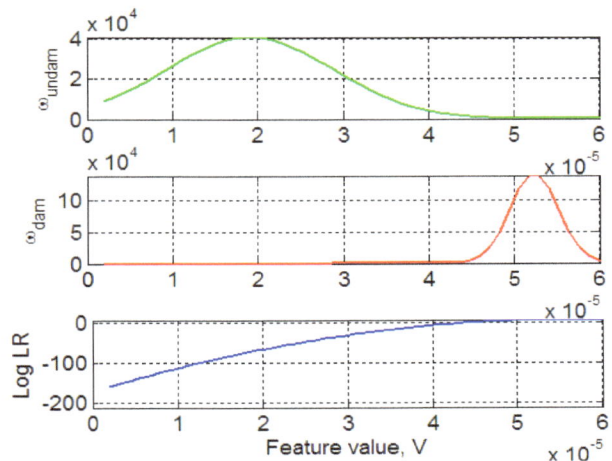

Figure 7. The Gaussian models of the probability density functions for classes "undamaged" (top), "damaged" (middle) and logarithm likelihood ratio (bottom).

(a) Undamaged

(b) Damaged

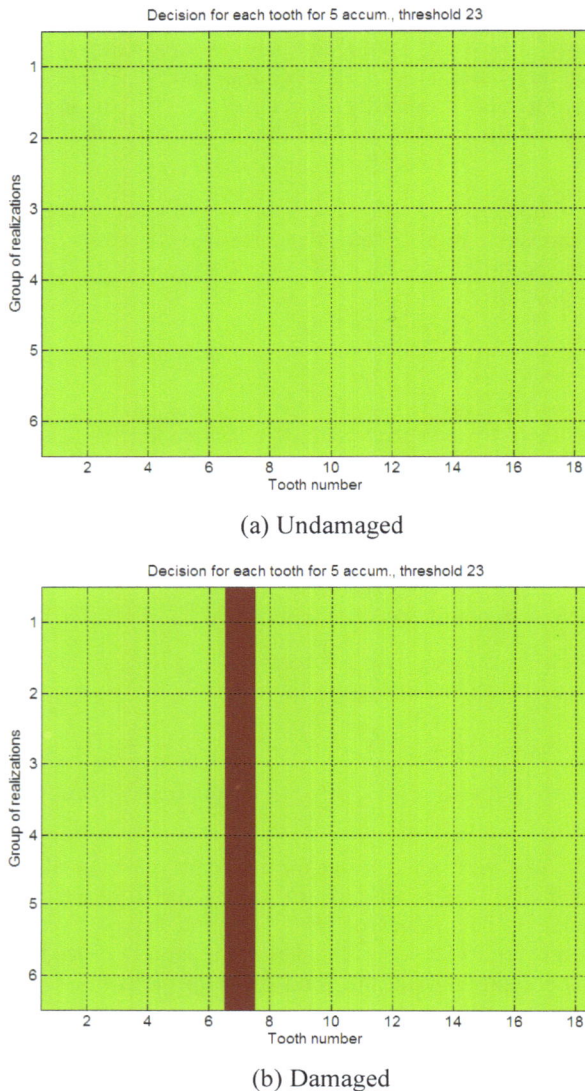

Figure 8. Diagnosis results for undamaged (a) and damaged (b) gears. The green colour is for decision "undamaged", the red colour is for decision "damaged".

The performed diagnosis has shown no false alarms for the undamaged pinion and stable damage diagnosis of tooth 7 for the damaged pinion (Fig. 8).

6. CONCLUSIONS

The wavelet technology for gear fatigue damage diagnosis was successfully applied for the first time for diagnostics of the partly missing tooth. Experimental investigation was performed using the Machine Fault Simulator.

The angular re-sampling of raw vibration data is performed automatically using the mesh components of the vibration data. The residual signal is obtained from the TSA signal by removing not only mesh harmonics but also low order shaft harmonics.

The main parameters of the technology were optimized using the total error probability as the optimization criterion. The missing tooth has been detected and continuously diagnosed throughout the whole experimental test without false alarms and missed detections using the advanced decision making technique based on the likelihood ratio. The likelihood ratio was obtained using the Gaussian models of the data for classes "undamaged" and "damaged". The time required for a single damage diagnosis was equal to 1 min 40s which is equivalent to 5 realizations of the residual signal.

Investigation of the technology effectiveness at different shaft speeds, loads, sampling frequencies and for gears with multiple teeth damages is planned in the near future. The possibility to use historical data from the gear with damage for technology calibration is also considered for a future work.

REFERENCES

Brie, D., Tomczak, M., Oehlmann, H., & Richard, A. (1997). Gear Crack Detection by Adaptive Amplitude and Phase Modulation. *Mechanical Systems and Signal Processing*, vol. 11(1), pp. 149–167.

Choy, F. K., Polyshchuk, V., Zakrajsek, J. J., Handschuh, R. F., & Townsend, D. P. (1996). Analysis of the Effects of Surface Pitting and Wear on the Vibration of a Gear Transmission System. *Tribology International*, vol. 29(1), pp. 77–83.

Combet, F., & Gelman, L. (2007). An Automated Methodology for Performing Time-Synchronous Averaging of a Gearbox Signal Without Speed Sensor. *Mechanical Systems and Signal Processing*, vol. 21(6), pp. 2590-2606.

Combet, F., & Gelman, L. (2009). Optimal Filtering of Gear Signals for Early Damage Detection Based on the Spectral Kurtosis. *Mechanical Systems and Signal Processing*, vol. 23(3), pp. 652–668.

Combet, F., Gelman, L., Anuzis, P., & Slater, R. (2009). Vibration detection of local gear damage by advanced demodulation and residual techniques. *Proc. IMechE, vol. 223 Part G: J. Aerospace Engineering*, pp.507-514.

Combet, F., Gelman, L., & LaPayne, G. (2012). Novel Detection of Local Tooth Damage in Gears by the Wavelet Bicoherence. *Mechanical Systems and Signal Processing*, vol. 26, pp. 218-228.

Dalpiaz, G., Rivola, A., & Rubini, R. (2000). Effectiveness and Sensitivity of Vibration Processing Techniques for Local Fault Detection in Gears. *Mechanical Systems and Signal Processing*, vol. 14(3), pp. 387–412.

Endo, H., & Randall, R. B. (2007). Enhancement of Autoregressive Model Based Gear Tooth Fault Detection Technique by the Use of Minimum Entropy Deconvolution Filter. *Mechanical Systems and Signal Processing*, vol. 21(2), pp. 906–919.

Feng, Z., Liang, M., Zhang, Y. & Hou, S. (2012). Fault Diagnosis for Wind Turbine Planetary Gearboxes via Demodulation Analysis Based on Ensemble Empirical Mode Decomposition and Energy Separation. *Renewable Energy*, vol. 47, pp. 112-126.

Forrester, B. D. (1996). *Advanced Vibration Analysis Techniques for Fault Detection and Diagnosis in Geared Transmission Systems*. Ph. D. dissertation. Swinburne University of Technology, Melbourne, Australia.

Gelman, L., Zimroz, R., Birkel, J., Leigh-Firbank, H., Simms, D., Waterland, B., & Whitehurst, G. (2005). Adaptive Vibration Condition Monitoring Technology for Local Tooth Damage in Gearboxes. *Insight Int. J. Non-Destructive Testing and Condition Monitoring*, vol. 47(8), pp. 461–464.

Gryllias, K. C., Gelman, L., Shaw, B. & Vaidhianathasamy, M. (2010). Local Damage Diagnosis in Gearboxes Using Novel Wavelet Technology. *Insight*, vol. 52(8), pp. 437-441.

Halima, E. B., Shoukat Choudhuryb, M. A. A., Shaha, S. L., & Zuoc, M. J. (2008). Time Domain Averaging Across all Scales: a Novel Method for Detection of Gearbox Faults. *Mechanical Systems and Signal Processing*, vol.22(2), pp.261–278.

Jiang, L., Liu, Y., Li, X. & Tang, S. (2011). Using Bispectral Distribution as a Feature for Rotating Machinery Fault Diagnosis. *Measurement*, vol. 44, pp. 1284-1292.

Lebold, M., McClintic, K., Campbell, R., Byington, C., & Maynard, K. (2000). Review of Vibration Analysis Methods for Gearbox Diagnostics and Prognostics. *Proc. of the 54th Meeting of the Society for Machinery Failure Prevention Technology*, Virginia Beach, VA, May 1-4, 2000, p. 623-634.

Lee, J. Y., & Nandi, A. K. (2000). Extraction of Impacting Signals Using Blind Deconvolution. *Journal of Sound and Vibration*, vol. 232(5), pp. 945–962.

Lee, S. K., & White, P. R. (1998). The Enhancement of Impulsive Noise and Vibration Signals for Fault Detection in Rotating and Reciprocating Machinery. *Journal of Sound and Vibration*, vol. 217(3), pp. 485–505.

Lei, Y., Zuo, M. J., He, Z., & Zi, Y. (2010). A multidimensional hybrid intelligent method for gear fault diagnosis. *Expert Systems with Applications*, vol. 37 (2), pp. 1419–1430.

Lin, J., & Zuo, M. J. (2003). Gearbox Fault Diagnosis Using Adaptive Wavelet Filter. *Mechanical Systems and Signal Processing*, vol. 17(6), pp. 1259–1269.

Loutridis, S. J. (2006). Instantaneous Energy Density as a Feature for Gear Fault Detection. *Mechanical Systems and Signal Processing*, vol. 20(5), pp. 1239–1253.

Maynard, K. P. (1999). Interstitial Processing: The Application of Noise Processing to Gear Fault Detection. *International Conference on Condition Monitoring*, University of Wales Swansea, April 12-15, pp. 77-86.

McFadden, P. D. (1986). Detecting Fatigue Cracks in Gears by Amplitude and Phase Demodulation of the Meshing Vibration. *Journal of Vibration, Acoustics, Stress, and Reliability in Design*, vol. 108, pp. 165-170.

McFadden, P. D. (1987). Examination of a Technique for the Early Detection of Failure in Gears by Signal Processing of the Time Domain Average of the Meshing Vibration. *Mechanical Systems and Signal Processing*, vol. 1(2), pp. 173–183.

Stewart, R. M. (1977). Some Useful Data Analysis Techniques for Gearbox Diagnostics. *Institute of Sound and Vibration Research*, Paper MHM/R/10/77.

Wang, W. Q., Ismail, F., & Golnaraghi, M. F. (2001). Assessment of Gear Damage Monitoring Techniques Using Vibration Measurements. *Mechanical Systems and Signal Processing*, vol. 15(5), pp. 905–922.

Wang, W. J., & McFadden, P. D. (1993). Early Detection of Gear Failure by Vibration Analysis—I. Calculation of the Time-Frequency Distribution. *Mechanical Systems and Signal Processing*, vol. 7(3), pp. 193–203.

Wang, W. J. & McFadden, P. D. (1996). Application of Wavelets to Gearbox Vibration Signals for Fault Detection. *Journal of Sound and Vibration*, vol. 192 (5), pp. 927-939.

Wang, W., & Wong, A. K. (2002). Autoregressive Model-Based Gear Fault Diagnosis. *ASME Journal of Vibration and Acoustics*, vol. 124, pp. 172–179.

Webb, A. (1999). Statistical pattern recognition. London: Arnold.

Yan, B. F., Miyamoto, A. & Brühwiler, E. (2006). Wavelet Transform-Based Modal Parameter Identification Considering Uncertainty, *Journal of Sound and Vibration*, vol. 291, pp. 285–301.

Zamanian, A. H. & Ohadi, A. (2011). Gear Fault Diagnosis Based on Gaussian Correlation of Vibrations Signals and Wavelet Coefficients. *Applied Soft Computing*, vol. 11, pp. 4807-4819.

Role of Prognostics in Support of Integrated Risk-based Engineering in Nuclear Power Plant Safety

P.V. Varde[1,2] and Michael G. Pecht[1]

[1]Center for Life Cycle Engineering
University of Maryland, College Park, MD, USA, 20742

[2]Life Cycle Reliability Engineering Lab
Safety Evaluation and MTD Section
Research Reactor Services Division
Bhabha Atomic Research Centre, Mumbai, India 400 085
varde@barc.gov.in

ABSTRACT

There is a growing trend in applying a prognostics and health management approach to engineering systems in general and space and aviation systems in particular. This paper reviews the role of prognostics and health management approach in support of integrated risk-based applications to nuclear power plants. The review involves a survey of the state-of-art technologies in prognostics and health management and an exploration of its role in support of integrated risk-based engineering and how the technology can be adopted to realize enhanced safety and operational performance. An integrated risk-based engineering framework for nuclear power plants has been proposed, where probabilistic risk assessment plays the role of identification, prioritization and optimization of systems, structures, and components, while deterministic assessment is performed using a prognostics and health management approach. Keeping in view the requirements of structural reliability assessment, the paper also proposes essential features of a 'Mechanics-of-Failure' approach in support of integrated risk-based engineering. The performance criteria used in prognostics and health management has been adopted to meet requirements of risk-based applications.

1. INTRODUCTION

Nuclear power, with over 430 nuclear power plants (NPPs) operating around the world, is the source of about 17% of the world's electricity. The nuclear industry has arrived at a point where it is dealing with two major issues. First, addressing the life extension for legacy units while complying with present day safety regulations. Second, designing new systems with enhanced safety features so that the core damage frequency meets the target of 10^{-6}/reactor years or less. The literature available suggests an increasing

role for a risk-based (RB) / risk-informed (RI) approach to the design, operation, and regulation of nuclear power plants in order to improve safety (IAEA, 2010; IAEA, 1993, 2010; Kadak and Matsuo, 2007).

This paper presents a role for prognostics - a relatively new paradigm, as part of risk-based approach to extend the present activities of monitoring, surveillance, in-service inspection, and maintenance from the periodic to condition-based through the application of prognostics methods.

The major elements of prognostics are online monitoring of precursor parameters and the detection of deviation from the reference condition using prognostic algorithms (Pecht, 2008). Here, the evaluation of remaining useful life (RUL) for the monitored component or system and the use of insights from this evaluation is a crucial part of risk-based / risk informed applications (Coble and Hines, 2010) Figure 1 shows the major steps in prognostics as part of integrated risk-based engineering (IRBE) applications. The main aim here is to monitor the degradation in a dynamic manner and enable prediction of the failure well in advance so that failure can be avoided altogether or advance action can be taken to repair or mitigate the consequences associated with the failure.

Traditionally, the nuclear industry has employed online status monitoring of safety and process parameters so that any deviation from the reference operating conditions can be detected in time and, if necessary, automatic safety actions can be initiated. Also, there exist various levels of defense in the form of alternate provisions that provide coping time for systems and equipment should the preceding level of defense fail.

However, there is a need to predict the life and reliability of each level of defense in order to enhance the safety of a plant. The prognostic approach facilitates the health management of systems and components based on the remaining life prediction of components. Even though in the current generation of plants, prognostic principles are used in the form of qualitative reliability and life attributes, the

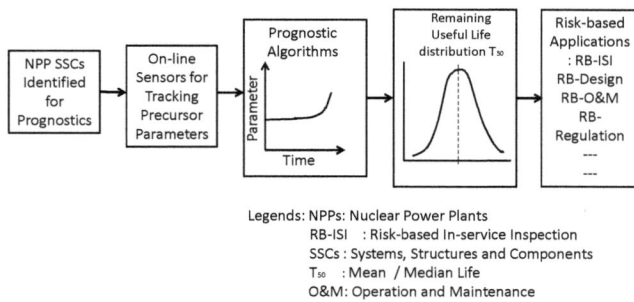

Legends: NPPs: Nuclear Power Plants
RB-ISI : Risk-based In-service Inspection
SSCs : Systems, Structures and Components
T_{50} : Mean / Median Life
O&M: Operation and Maintenance

Figure 1. Simplified representation of prognostics as part of integrated risk-based application.

full potential of prognostics has yet to be realized through the formal implementation of a prognostics-based health management program.

Although the healthcare industry provides inspiration and ideas related to the development of prognostics for engineering systems, the available literature shows that the role of prognostics is growing in many fields of components and systems where safety forms the bottom line, such as in aerospace (Wheeler, Kurtpglu and Poll, 2010), electronics systems (Kalgren et.al.2010; Bhambra, J.K, 2000; Mishra et. al., 2004)), telecommunications, and structural systems (Guan, Liu, Jha, Saxena, Celaya, and Geobel, 2011). Specific engineering applications include prognostics for bearings and gears (Klein, Rudyk, Masad, and Issacharoff, 2011), engine/turbine condition monitoring (Wu, 2011; Hyres, 2006), aircraft engine damage modeling (Saxena, 2008), aircraft ac generator model simulation (Tantawy, Koutsoukos, Biswas, 2008), health monitoring of lithium ion batteries (Chen and Pecht, 2012), and development of an intelligent approach in support of diagnostics and prognosis (Chen, Brown, Sconyers, Zhang, Vachtsevanos and Orchard, 2012). Based on the experience in these fields and the knowledge that has been generated over the years, it can be argued that a prognostics-based approach, as an extension of a condition-monitoring approach, is expected to go a long way to address the surveillance and monitoring requirements of new as well as old nuclear plants. For old plants, the life extension program can be implemented on a sound footing by integrating prognostics and health management models to complement the risk-based approach. For new systems, enhanced safety can be achieved by the implementation of prognostics-based health management of systems and components. To realize risk reduction through the prognostic approach, design specifications should ensure that a plant is built with online monitoring capabilities for the identified precursor parameters. This basic setup will focus on online prediction of remaining life and reliability considering the postulated loads and stresses such that risk reduction by detecting failure in advance can be realized. The same approach applies to legacy plants also. In these plants, the existing sensors and monitoring systems can be adopted in support

of prognostics, like the vibration monitoring data on rotating machines can be utilized for prognostics and health management. It is relatively easy to install a vibration monitoring network for existing check-valves. However, there will be issues related implementing on-line monitoring for some specific locations (e.g. for in-core and reactor support and structural components which may not be easily accessible). For these systems the monitoring of derived or secondary parameters may work. For example the annulus gas monitoring system provides information of leakage, if any, as an on-line assessment for integrity of coolant channel in the existing fleet of Pressurized Heavy Water Reactors. Periodic inspection, installation of coupons (e.g. to assess corrosion of sub-soil piping) may also provide effective approach in the absence of on-line monitoring for existing plants.

This paper presents a review of the current approaches to monitoring and surveillance. We assess the prognostic requirements as a part of an integrated risk-based approach for old and new NPPs. Even though the implementation of prognostics varies depending on the type of components and the objective of prognostic applications, this paper emphasizes proposing a general framework that addresses the basic or broader aspects of various applications. The prognostic performance metrics and other related issues that are relevant to NPPs are also discussed.

2. SURVEILLANCE AND CONDITION MONITORING IN NUCLEAR PLANTS: A BRIEF OVERVIEW

The design safety philosophy for nuclear plants requires the implementation of defense-in-depth, fail-safe criteria: the design is fault-tolerant to the extent that a single failure event will not adversely affect plant safety. The selection of online process parameters and associated limiting condition settings ensure the monitoring of all postulated conditions and the taking of timely action such that safety is not compromised.

Most of NPPs operating world over belong to first- and second-generation systems. Based on the accumulated operating time logged by the operating NPPs, the average life of the NPPs works out to be over 20 years (Bond, Doctor and Taylor, 2008; Bond, Tom, Steven, Doctor, Amy, Hull, Shah, and Malik, 2008). In general, NPPs have a design life of more than 40 years. The evidence of aging may manifest in many ways, like frequent failure of components in process and safety systems and subsequent interruption of plant operation, overall reduction in plant availability, adverse impact on available redundancy or safety margin in safety systems, etc. (IAEA, 1995; IAEA, 2009b). With effective inspection and maintenance practices, degradation due to age can be managed and operational life can be extended. For over 30 years the United States (U.S.) nuclear power industry and the U.S. Nuclear Regulatory Commission (USNRC) have worked together to develop aging management programs that ensure the plants can be operated safely well beyond their original design life (Gregor and Chokie, 2006).

Third generation plant designs are characterized by the use of inherent safety features such as negative void coefficient of reactivity, incorporation of passive features, the shift from analogue to digital plant protection systems, and added redundancy from 2-out-of-3 in second generation to 2-out-of-4 trains and channels (including the control and protection system and improved accident management features in containment). Application of the leak-before-break concept in design and operation has been associated with new plants. Apart from this, condition monitoring using vibration signatures, current signatures, insulation resistance assessments, temperature trends, acoustic signatures and other process parameter variations forms part of diagnostics and in a limited way prognostics assessment of the third generation plants components and systems. Some examples of condition monitoring include assessment of the health of the fuel by online monitoring of radiation level, assessment of rotating machine mechanical bearing condition based on online or off-line measurement of vibration and temperature, current signature analysis to assess the health of induction motors, electromagnetic interference mapping to assess the effect of magnetic field, pump shaft performance monitoring using eddy current technique, exhaust air temperature and smoke quality monitoring to assess health of the diesel generators, and oil sample analysis for foreign material to assess degradation and wearout of mechanical parts.

There are also examples of built-in-test (BIT) facilities for online diagnostics in systems and control systems. For safety channels, the protection channel will be activated only when there is demand. In these types of systems, the latent fault remains passive and reveals itself only when a channel is required to be activated. For such cases periodic testing is conducted to reveal a passive fault so that a system is available when there is an actual demand. However, the test interval determines system availability. The safety objective requires that the channel should be tested as frequently as possible to ensure the maximum availability of the channel. For a protection channel this testing is conducted by incorporating a fine impulse test (FIT) feature. An FIT module sends an electrical pulse of very short duration of around ~2 milliseconds. This duration is long enough to test electronic cards but short enough to not activate an actuation device, such as an electro-magnetic relay as actuation of a 48 VDC relay requires a signal that prevails at least for ~ 40 milliseconds.

From the structural health monitoring point of view, annulus gas monitoring, where CO_2 gas is passed between an annular gap between the pressure tube and a calandria tube, is a good example of condition monitoring (IAEA, 1998; Baskaran, 2000). The dew point of the CO_2 is monitored at

the exit point of the channel to identify any indications of leak. Any increase in dew point from the reference dew point of around -40°C indicates a possible leak in the annular region from the pressure tube or calandria tube and prompts an analysis of the region. This is an example of an implementation leak before the break strategy in real-time mode.

The examples listed above are not exhaustive. They indicate the state of the art in operating nuclear plants that have condition monitoring provisions and limited features for prognostics. However, this background provides a basis for identifying gap areas for the implementation of prognostics and health management program as part of a risk-based approach.

The first step in implementing a prognostic program for a complex system such as an NPP is to classify the systems, structures, and components (SSCs) into different categories, keeping in view the NPP's design and operation characteristics that will also determine the type and level of prognostics. The classifications and categorizations performed in this paper are not comprehensive but rather indicative. The objective here is to present the state of the art of monitoring for these components from the point of assessing the prognostic maturity level for these components. Table 1 shows the status of online monitoring, condition monitoring, in-service inspection, and diagnostic prognostics. This categorization has been primarily done keeping in mind the pressurized heavy water reactor systems and components and is representative and not exhaustive. The basic idea is to categorize the systems and components, as shown in Table 1, keeping in mind the reactor type and prognostic requirements. The following are points drawn from this table with respect to criteria required for the classification of NPP SSCs, the status of various surveillance methods, and existing gaps in the implementation of prognostics:

This classification is required for both new and old reactors. In fact this table provides a good starting point to generate prognostic specifications for the new design. In each category, the representative components are chosen such that they fulfill one or more of the following criteria: the component allows prognostic implementation (the most challenging), the component allows prognostic, diagnostic, in-service-inspection or condition monitoring implementation, the component represents a typical sample from the category, and that for other components in the group, prognostic implementation will be similar to the representative component. The monitoring program has matured for all the categories of components in NPPs. In-service-inspection (ISI) is applicable to mechanical components in general and piping and associated fittings in particular. It may be noted that the capability of ISI in terms of various coverage factors such as detection, location, and isolation remains a subject of research and development

Component / System type (*Representative items*)	M	OFL ISI	OFL CM	OFL D	OFL P	ONL D	ONL P	Remarks / (references)
a) **Reactor Structure:** Reactor Pressure Vessel, *Coolant Channels*, Reactor block, Reactor vault and its lining, Shielding structures, Steam Generator, and associated fittings and penetration and nozzles, and, Ventilation plenum and ducts etc.	***	****	***	***	****	***	**	The life prediction for Candu / PHWRs) coolant channels (pressure tube; IAEA, 1998; Dharmaraju, 2008; Chatterjee, 2012)
b) **Non-Reactor Structure:** *Containment* and civil structures, Fuel Transfer and Storage block, Overhead tanks and reservoirs, Airlocks, Structural support, RB Dampers, Bridges and jetties, guide and support etc	****	****	****	***	***	***	**	Structural health prediction in R&D stages.(Andonov, 2011; Coble, 2012),
c) **Mechanical Components:** *Pumps & Turbines, Piping,* Valves, Heat Exchangers (Shell and Tube and plate type, Fueling Machine, Fans and Dampers, Hydraulic drives and systems, Strainers and Filters, Bearings, Diesel Generators, Compressors, Cranes, Travelling water screens, etc	****	****	****	****	**	***	***	State of the art is available on on-line diagnostics. Prognostics in R&D stages, (Heng A, 2009; Samal, 2010; Coble, 2012)
d) **Electrical Power System:** Electrical buses and cables, HV Transformers, **Motors**, Breakers and Isolators, Power Relays, Motor Generator / alternator Sets, Battery banks etc.	****	***	***	***	***	**	**	CM for rotating machines. (Heng, 2009)
e) **Power Electronics systems:** *Un-interrupted Power Supplies,* Convertors, Invertors and rectifiers etc.	***	**	**	**	**	**	*	R&D work on Capacitor, IGBT reported. (Yin, 2008; Smith, 2009; Ye, et.al., 2006)
f) **Micro-electronic Systems:** *Digital Cards,* ICs, PLCs and FPGAs, interconnects and Control Cables, Control Connectors etc.	****	***	****	***	**	****	**	Prognostics in R&D stages (Pecht, 2008)
g) **Process Instrumentation:** Electrical and Pneumatic *transmitters,* Level, Pressure and Flow gauges, RTDs and Thermocouples, Impulse tubing, Control Valve telemetry, Solenoids, pH, Conductivity meters.	****	****	****	***	**	****	**	Smart sensors and periodic calibrations, (Hashemian, IAEA-CN-164-7S05)
h) **Nuclear instruments:** *Fission Counters,* Ion Chambers, etc.	****	****	****	***	**	****	**	Often saturation characteristics indicate reaming useful life.

Note: The characterization of the metrics has been done considering the 'representative items' identified in column with 'bold and italics'.

Legends: **M**: *Monitoring;* **CM**: *Condition Monitoring,* **D**: *Diagnostics;* **P**: *Prognosis;* **OFL**: *Off-line;* **ONL**: *Online; ISI: In-service-Inspection;*

'****': *Technology Available for NPPs;* '***': *Technology Available further qualifications are required for specific applications;*

'**': *Technology in R&D domain, feasibility demonstrated;* '*': *Work initiated; x : No work reported in literature.*

IMPORTANT: The items shown in table provide an overview and do not claim, in any way, to provide specifics/guidelines.

Table 1: Categorization of SSCs and Status of Monitoring, Diagnosis and Prognostics in Existing (up to Generation III NPPs)

(Coppe, Haftka, Kim, and Bes, 2008). The surveillance activities, which include testing and maintenance, performed on electronics channels and electrical power supply systems have also been categorized under the ISI program. Condition monitoring programs for reactor coolant channels, pumps, motors, bearings, reactor containment, fission counters, and transmitters is mature in NPPs.

Generally, diagnostics is provided for selected components, such as diesel generators, pumps, and digital cards. For example, the complete protection channel is monitored in an online mode for detecting failure of any card using a built-in-test or a fine impulse test facility.

There are many examples of online surveillance and health management programs in NPPs. Some of the examples include coolant channel inspection activities (item (a) in Table 1) in a pressurized water reactor or an advanced aging management program for mechanical components. Similarly, the fine impulse test facility for monitoring all the redundant channels (item (f) in Table 1) is also an example of a Verification and Validation (V&V) tool for the health assessment of electronic parts of protection channels. The saturation characteristics of ion chambers or fission counters provide an online indication of the remaining life of these components. However, regulation and protection channels only have diagnostic features, and research and development is required for the implementation of prognostics for these components.

At the component level, the condition monitoring of rotating machines using vibration and temperature monitoring and diesel generators sets are arguably a mature health management program, except that they lack the capability of life prediction. Our literature search suggests that online prognostics either have not been developed or still in a research and development stage for most of the components in NPPs.

3. INTEGRATED RISK-BASED APPROACH

The term 'risk' in nuclear parlance deals with assessing the likelihood and consequences for a given scenario. When the modeling is performed considering risk as the major objective metric by integrating probabilistic and deterministic methods, then the approach is called as integrated risk-based engineering (IRBE). Even though the majority of risk modeling is performed considering hardware failures, incorporation of human factor modeling into plant modeling also forms a significant feature of risk models. Another major feature of this approach is that it provides a quantified statement of safety (Tsu-Mu, 2007). Deterministic criteria, design, and operation information form part of the risk assessment to reflect a realistic representation of the plant model. There are many approaches to risk assessment, including hazard and operability analysis, failure modes and effects analysis (FEA), what-if approaches, cause/consequence analysis, and

quantitative methods for nuclear plants. The probabilistic risk assessment (PRA) methodology is a well-accepted methodology for risk assessment of the plant. Apart from this, probabilistic interference modeling using stress-strength distribution approaches is also used for determining failure criteria at the component level. The following section provides a brief discussion on PRA, as this approach handles the probabilistic element of the integrated risk-based approach.

3.1. PROBABILISTIC RISK ASSESSMENT

Probabilistic risk assessment (PRA) is an analytical approach to predicting the potential off-site radiological consequences of accidents for a nuclear power plant and research reactor. PRA is performed at three levels. Level 1 PRA deals with system modeling to provide estimates of core damage frequency. Level 2 PRA deals with the release mode and mechanisms from reactor containment to provide estimates of radioactivity release frequencies for various source terms. In level 3 PRA consequences for various releases are estimated to provide an assessment of risk. As level 1 PRA deal with SSCs modeling level 1 PRA is directly relevant to prognostics application. Hence, for prognostic applications, results and insights from level 1 PRA with assessments of system-level unavailability and core damage frequency as indicators of safety are proposed in this paper.

The major elements of level 1 PRA methodology include: identification of the postulated initiating event (both internal and external), assessment of the response of the plant using event tree methodology, modeling for safety system failure employing a fault tree approach, quantification of the model by using failure data including human error probabilities and common cause failure data, etc., iterative simulation of an integrated plant model to estimate the core damage frequency and uncertainty bounds, and sensitivity analysis for critical assumptions made during the study. Figure 2 shows the general level 1 PRA methodology.

Risk-based applications, such as an equipment surveillance test interval and allowable outage time optimization, precursor event identification and analysis, evaluation of emergency operating procedures, and risk monitoring for plant configurations studies are generally based on level 1 PRA studies. The reason for this is that level 1 PRA deals with modeling plant configurations with consideration of component failures, human actions, test and maintenance data, and operational procedures and plant technical specifications to predict unavailability and core damage frequency at system level and plant level. Accordingly, we deal with level 1 PRA, wherein the estimates of core damage frequency represents a statement of risk.

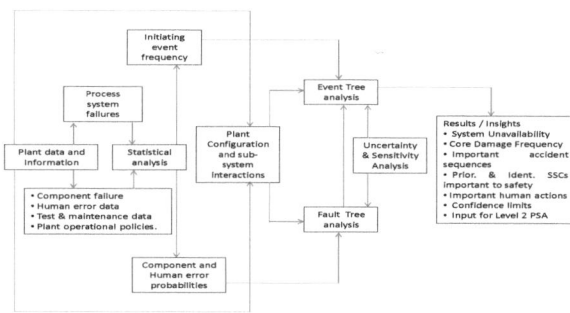

Figure 2. Level 1 PRA methodology.

PRA methodology can be considered mature enough to be used in decision support. The main reasons for the growth of this field are because it enables the quantification of the statement of plant safety by estimating the core damage frequency and unavailability at the system level, it provides a mechanism to capture random elements of safety characterizing uncertainty in safety and performance estimates, it provides a strong framework for integrated performance of SSCs into the model of the plant, and it facilitates integration of the human factor into an integrated model of the plant. When the PRA framework is employed in support of regulatory decisions, then the application is called "risk-informed."

3.2 Role of PRA in Risk-based Applications

PRA provides a systematic framework for the identification of safety and reliability issues, safety-based prioritization of components, human actions, and assessment of plant configuration. The major result of PRA is a statement of core damage frequency per year, while the assessment of design strength and weaknesses of the plant, characterization of uncertainty, and sensitivity analysis also form part of the major insights.

The level 1 PRA model and framework is used for many risk-based applications, including risk-based design optimization, risk-based in-service inspection, risk-based maintenance management, risk-based technical optimization, risk monitoring, and precursor analysis. Among these, the risk monitors are used to address real-time issues and have gradually become an integral part of the operation of NPPs in many countries. This is the reason why PRA applications are becoming an integral part of regulatory review as part of a risk-informed approach (Tsu-Mu, 2007).

3.3 Living PRA and Risk Monitor

Even though the traditional approach to PRA modeling is static in nature, the application of PRA as 'living PRA' and 'risk monitor' make this approach in a limited sense dynamic in nature. The living PRA approach ensures updating of the plant PRA model on a periodic basis such

that it reflects the as-built and as-operated features of the plant. These living PRA models are updated based on modifications or change in operating procedures or regulatory stipulations. Changes are documented in such a way that each aspect of the model can be directly related to existing plant information, technical specifications, and emergency and normal operating procedures. The verification of assumptions within the analysis and the associated sensitivity analysis forms part of living PRA approach (IAEA, 1999).

There is a noticeable growth in on-line application of PRA as risk monitor. These risk monitors provide assessment of risk for real-time changes in equipment configurations and technical specification parameters like allowable outage time or change in test intervals, etc. Risk monitors reflect the current plant configuration in terms of status of the various systems and components. Basically, risk monitors are developed as an operator aid in decision-making support in the plant control room environment. The operator may like to assess change in risk levels for an action involving, for example, taking any components out of service for maintenance or tests. Given the above background, living PRA and risk monitors make the risk assessment process dynamic in a limited sense. This means that it addresses discrete events/changes in time not in the continuum sense of time. The risk monitor models should be consistent with living PRA models and should be updated at least with the same frequency as living PRA models (NEA/CSNI, 2005). The change in core damage frequency or core damage probability for a given scenario assessment and its comparison with the quantitative criteria is used in support of decisions as part of risk-informed approach.

3.4 Limitations of the Risk-based Approach

The risk-based framework in its present form is essentially 'static' in nature and often incapable of conducting evaluations involving dynamic scenarios evolving in through time. Hence, there is a need to make the whole approach more dynamic in nature for addressing real-time scenarios. It must also account for degradation, which is inherent in systems and components, and have a predictive capability with reasonable accuracy such that it can provide a time window for corrective actions. It should also have risk mitigation or management features. This is where a prognostics approach can enable evaluation of dynamic scenarios. The probabilistic tools and methods, on the other hand, can provide the required framework for assessment of available safety margins and characterization of uncertainty. This approach has been extensively applied to mechanical and structural systems, like risk-based in-service inspection of piping and structural systems, risk-based maintenance for process systems, etc. However, radiation-induced degradation poses a challenge to life prediction of structural components in NPPs. There are some areas in NPPs where radiation-induced degradation modeling and assessment

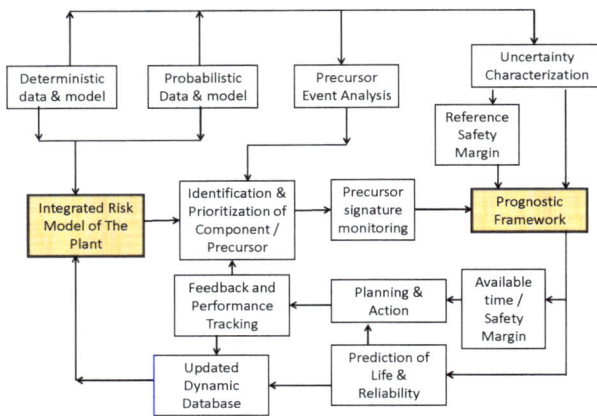

Fig. 3 Interrelationship of IRBE and prognostics approach.

have been performed to predict the remaining life of structural systems, such as pressure tube life prediction in PHWRs and CANDU reactors, reactor vessel health assessment modeling, and aging of control and power cables. However, these pose a challenge to the assessment of life of in-core components. Extensive research and development is being performed to implement this approach for micro-electronic and power-electronic systems (Tsu-Mu, 2007; Patil, et. al., 2009).

4. PHM and IRBE

Figure 3 shows the interrelationship between integrated risk-based engineering (IRBE) and a prognostics framework. As can be seen, the deterministic and probabilistic methods together can be used to build an integrated risk model of the plant. The integrated risk model provides information about safety issues in general, which is vital for the implementation of prognostics. This includes information on components and modes and sequences of system failure, which are precursors that can form candidate components for implementation of prognostics. From here the organization gets vital input to focus on a small group of safety-critical items to improve technical specifications. This approach provides a valuable tool for prognostic coverage so that maximum benefits can be realized by investing the available resources.

While probabilistic input is critical for uncertainty characterization in prognostics, the overall effect is to help in the assessment of a safety margin at the reference level. Once a prognostic assessment is performed, it is possible to understand the safety margin in a dynamic manner. The prognostic strategy complements the present risk-based framework by online performance tracking and generation of feedback for operators as well as regulators. The overall gain from the implementation of a prognostics strategy is that it allows the assessment of safety issues by predicting life, reliability, and prognostic distance in a dynamic manner.

5 REQUIREMENTS OF PROGNOSTICS FOR IRBE APPLICATIONS

There are three major areas that need to be strengthened in the risk-based approach. These include dynamic modeling, improved methods for uncertainty characterization, and realistic assessment of safety margins. It can be argued that a risk-based approach is a rational one, in that it combines the plant configuration, including operational logic, and performance parameters through probabilistic reasoning. Hence, this approach, in conjunction with the deterministic approach, is expected to provide more flexibility compared to the traditional approach. Apart from this, performance monitoring and generation of feedback following the implementation of changes and modifications forms an integral feature of risk-based engineering. Development and implementation of prognostic program is expected to address the above issues in the following manner:

Develop suitable sensors that can measure a precursor parameter of interest. Assess online the reliability and remaining useful life of the monitored systems based on identified precursors or degradation characteristics. Develop a prognostic algorithm that provides advance remaining life prediction with an adequate time window. Characterize uncertainty in the prediction, instead of point estimates of RUL, such that management issues can be addressed in an efficient and effective manner. Employ multi-objective algorithms that take into consideration risk, cost, and reduction in radiation dose. The prognostic framework should have provision for database, model-based (Physics of Failure (PoF) and Mechanics of Failure (MoF)), and fusion approaches, keeping in view the varying nature of prognostics programs for a range of components, such as mechanical and electrical systems, electronics, and nuclear components. The prognostic algorithm also should have a provision to provide feedback online to track the performance of modifications in a component that has been replaced or has undergone a maintenance procedure or calibration of some instrumentation.

It may be noted that use of input from prognostics may require modification to the existing risk assessment approach. For example, an existing database in the risk-based approach may have only static reliability data, failure criteria, and maintenance and test schedules. However, when the dynamic aspects are implemented as part of prognostic feedback to risk-based engineering, then there is a need to re-organize the complete database framework such that dynamic inputs and outputs can be managed.

5.1 Prognostics Design Requirements

The complex nature of NPP design requires a different set of design metrics that satisfies the requirements of a particular application. There are many areas that require research and development with respect of material degradation, development of special sensors, and suitable

Parameter	Applicable Levels				
	1	2	3	4	5
Plant – stage (* and #)	Under Design New	Operating plant (Useful Life)	Aged Operating plant	Shutdown Under-refurbishing	Operating After refurbishing
Objective (* and #)	Improve Safety	Improve Availability & Safety	Monitoring of remaining useful life	Follow-up after retrofitting	Reduction in Operational Cost
State-of-the-art enabler (*)	Online Monitoring, Off-line Diagnosis.	Online Monitoring Off-line Diagnosis and Conditioning Monitoring.	Online Monitoring, diagnosis and condition monitoring.	Online Monitoring, diagnosis and condition monitoring, off-line prognosis.	Online Monitoring, online diagnosis and condition monitoring, online prognosis.
Subject (* and #)	Micro-Electronics / Digital Control Channels	Power Electronic	Electrical	Structural / Mechanical	Interdisciplinary
Implementation Level (* and #)	Level 1	Level 2	Level 3	Level 4	-
Risk Assessment Approach (* and #)	Qualitative / Deterministic	Failure Mode Effect and Criticality Analysis	Hazard and Operability Analysis	System Level Reliability Modeling Fault tree. Event Tree	Level 1 Probabilistic Risk Assessment (L-1)
Existing Maintenance /Health Management Strategy (*)	Preventive Maintenance and Scheduled Testing	Condition based test and Maintenance	Reliability Centered	Risk-based Test and Maintenance	Online reconfiguration
Stakeholders (* and #)	Design Team	Operating organization	Regulators	-	-
Implementation Approach (* and #)	Model Based	Data Based	Risk-based	Fusion	-
Availability of Tools and Methods / Challenges (* and #)	Prognostic algorithm	Availability of Sensors	Degradation Models	Feature Extraction Methods	-
Cost-benefits (* and #)	In the context of NPPs, cost benefits for a particular level is assessed using safety and availability indicators.				

Legend *: applicable to old plants; # : applicable to new plants

Table 2.0 Prognostics Design Requirement

algorithms for online feature extraction and analysis. In some situations, the design constraints may make it challenging to implement a prognostic program. Table 2 shows the requirements that can be used for the design of a prognostics program for a given application. The applicability of these parameters for old or new plants is indicated in the table by * and #, respectively.

5.1.1 Plant Stage

The scope of a prognostic program will be governed by such factors as at what stage of the plant the prognostics is being implemented. As mentioned earlier, for the implementation of a prognostic program in new plants, prognostic requirements and specifications should be part of the design strategy. Since the plant has yet to be built, provisions can be made in advance, keeping in view criteria such as safety and availability. Prognostics as part of life extension will have activities focused on select systems, components, and structures. However, the plant's design and operational constraints will dictate the implementation levels. Prognostic requirements for refurbished plants will be similar to a plant whose life has been extended. In a refurbished plant prognostics is useful particularly for those systems where clear insights into the remaining life of certain components is not available, while the cost of bulk replacement would have been prohibitive, and where it is felt that online prognosis and diagnosis would be useful, such as in coolant channels, piping, and power supply cables.

5.1.2 Objective

The objective of prognostics is defined as keeping in view the plant status, logistics, and data and knowledge base, particularly the understanding of the material degradation phenomenon and the availability of prognostic algorithms. For new plants, the institutive reaction will be to go for a model-based approach, while for older plant, where enough data are available, the data-driven approach will be preferred. It should be noted that most of the nuclear plants in the world either have a level 1 PRA model with internal initiating events for full power conditions or a reactor core as the source of radioactivity. These plants have an obvious advantage over using risk modeling to formulate or identify and prioritize a prognostic program. The available literature shows that most prognostic implementations produce improvement in availability and cost-benefit objectives (Hyers et al. 2006). There have been applications of prognostics to aircraft health monitoring, where the emphasis has been on safety. There have also been applications of PHM with mission safety as a driving force.

5.1.3 State of the Art

The state of the art refers to the current status of monitoring and surveillance methods used in a plant. If a plant is in the design stage, metrics may include the requirements for monitoring and health management. Provisions can then be made throughout the design stage for implementation of prognostic program. However, in the operating plants the existing sensors/provisions, and new sensors will determine the level and scope of prognostic. Generally, the state of the art in the current generation of plants facilitates the online monitoring of important safety parameters and condition monitoring and surveillance. Such programs are not fully automated, however; the diagnosis is performed in off-line mode for most systems. There are closed feedback loops wherein corrective actions are automatic. These feedback loops ensure the maintenance of plant parameters within set limits. These metrics help to determine the specifications and the scope of the prognostic requirements. The available literature on NPPs does not appear to provide information on application of prognostic based on online life and reliability prediction.

5.1.4 Subject

For complex systems such as NPPs most of the systems require an interdisciplinary approach to implement prognostic program. However, particular disciplines may require a unique focus. For example, prognostics for reactor protection channels and structural components such as reactor blocks will differ with respect to degradation mechanisms, the time window available, and the monitoring and sensor requirements. Hence, even though the broad framework may remain the same, the specifics will vary by applications.

5.1.5 Level of Implementation

The level of implementation metric is derived/adopted from the procedure developed by Gu et al., keeping in mind the NPP requirements. In this reference, various levels have been presented for the implementation of prognostics of electronic systems. For complex systems such as NPPs, there are different prognostic levels. Level 0 is the component level, which includes items such as fuel assembly, feeder, bearing, motor, control and power cables, alternator, pipeline, battery, relay, micro-processor chip, switch, and electronic cards. Level 1 includes the assembly of components of a particular class, such as mechanical, electrical, electrical, or nuclear and associated connections that perform a basic function. Examples include pumps with connected piping up to suction and discharge and the suction strainer, a compressor with sub-components such as coolers and associated connections, diesel generators with support systems, and power supply modules, amplifier modules, and function generators. Level 2 includes those systems that are activated only on demand from the plant control system. They can also be referred to as safety support systems such as class III electrical power supply, class II control supply, and class I power supply systems. Level 3 systems include those systems that are required to be operational when the reactor is in operational state, including the main coolant system, class IV power supply systems, feed water systems, regulation systems, and process water systems. The structural systems, such as the reactor vessel and reactor shielding components, reactor pile, and containment building, that are basically passive in nature but require a structural approach for health monitoring are categorized as level 4. These categories are based on the broad characterization of component functional requirements and their place in a system, whether as an independent unit, sub-block, block, major function, or assembly of functions to deliver an objective function.

5.1.6 Risk Assessment Approach

Most NPPs have a level 1 PRA implemented, considering the internal initiating of events for full power conditions. Even though risk-based applications require shutdown or a low power operation PRA, the availability of a full-power PRA can be considered for initiating a PHM implementation program. Apart from this, Failure Modes, Effects, and Criticality Analysis (FMECA) forms an integral part of PHM implementation. It is recommended that a comprehensive FMECA program should be initiated, keeping in mind the focus of prognostic implementation.

5.1.7 Existing Maintenance Health Management Strategies

This metric determines the current maintenance strategy, an important reference for building a PHM program. Typically, most nuclear plants use preventive maintenance as the major approach for health management. However, condition monitoring, in-service inspection, and scheduled test and maintenance are the general features for health management. The available literature shows that in some NPPs and industrial systems, reliability-centered maintenance, risk-based in-service inspection, and risk-based technical specification optimizations are also used (IAEA, 1993). The available framework is important, as the data generated on the maintenance and health of these systems and pieces of equipment form the fundamental part of the data-driven approach for prognostics. Along with inputs from risk models, these data and insights will help to identify and prioritize the prognostic program.

5.1.8 Stakeholders

Though stakeholders are not a metric, they affect which agency is interested in prognostic applications. The designers would like to have a prognostic program for identified systems or as part of a design policy for systems that they feel will determine the life of the plant. These could be in-core components or structures that form an integral part of systems such as reactor vessels, pile blocks, storage pool linings, or containment, or it could be some safety or process system for which it is important to track performance. For operational agency, it could be certain aspects of the plant that affect plant availability, such as performance of the strainer, check valves, and certain pipelines and bearings, which require continuous monitoring and remaining life assessment such that repair

and replacement of these components can be scheduled to improve plant availability. Regulatory agencies want to track the performance of a system where the changes have been implemented. Here the role of prognostics is to provide feedback on the remaining useful life or performance monitoring of a system for a specified period of time or for an extended duration to ensure that safety has not been compromised.

5.1.9 Approach for Implementation

The approach to prognostics implementation is governed by many factors, including the objective or purpose of prognostics, the level of detail required the availability of data, and plant constraints. For argument's sake, if a prognostic program requires performance monitoring as part of a risk-informed or risk-based approach, then the focus will be on monitoring the performance metrics of the system under regulatory review. If a prognostics program is being designed for a new plant where the objective is to strengthen the safety function, then the task should include the prognostic specifications in the design phase and keep provisions not only for online monitoring but also for the implementation and management of the health of the plant. If prognostics is being implemented as part of a life extension strategy, then it must be noted that the focus should be on structural remaining life assessment. As a rule, NPPs require close monitoring of structural health, particularly where safety is the major metric, even if it has not entered the aging phase.

5.1.10 Tools and Methods

Prognostic tools and methods are identified only when FMECA has been performed, precursors have been identified, and the broad approaches, including data-driven and PoF-based, have been evaluated, keeping in mind the requirements of applications. However, detailed studies, literature searches, or required meetings with consultants may present issues associated with selection of prognostic algorithms, the availability of sensors, the availability or limitations of degradation models, and approaches that will be required for feature extraction, deriving useful data and information from a host of complex data and signatures collected from experiments.

5.1.11 Cost-benefit Studies

The available literature shows that cost-benefit evaluation can be used to demonstrate the net benefit of the implementation of PHM results (Wood and Goodman, 2006). In the context of nuclear plants, benefits need not be in terms of monitory gain; they could be in terms of safety improvement, life extension, or lessening the burden on the operating staff.

4. PROGNOSTIC FRAMEWORK FOR NUCLEAR PLANTS

Major elements of prognostics implementation include monitoring system performance through noise analysis,

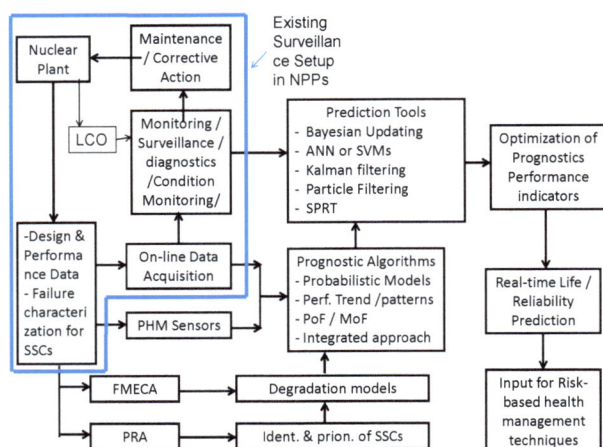

Legend: PRA: Probabilistic Risk Assessment; FMECA: Failure Modes, Effects, and Criticality Analysis; SSCs: Systems Structures and Components; LCO: Limiting Condition for Operations; SPRT: Sequential Probability Ratio Test.

Figure 4. Prognostics framework for NPPs.

detecting changes by trending, understanding and identifying root causes of failure, prognostics, and health management (Vichare and Pecht, 2006). Even though there will be variation in the tools and methods, the level of accuracy required when the prognostics is implemented for mechanical, structural, or electrical systems, will remain the same.

6.1 Existing Set-up

Traditionally, online monitoring and maintenance, including surveillance of passive and structural systems, as well as maintenance of active components form an integral part of NPP operations, as shown in Figure 4. The existing approach is shown within the boundary drawn on left side of the Fig. 4. The monitoring provisions exist at the component, system, and plant level. These monitoring provisions are limited to process parameter values and an equipment status display, which indicates various states of reactor operation including transient states. However, condition monitoring and surveillance for many systems is performed in an off-line mode as part of plant policy for selected equipment. Even though condition-monitoring approaches have matured and are being used in the health management, the prognostic quotient in terms of the prediction of remaining life is low. One of the major reasons for this is complexity in terms of material characterization, such as irradiation-induced degradation of core components and structures (Bond et al., 2008). Apart from this, the nuclear industry operates on conservative criteria; hence, strict regulations for design and operation dictate that uncertainty in real-time assessment should be as low as possible. However, in the present situation, advances made in other application areas (such as space, aircraft, and civil) can be implemented in NPPs by incorporating

adequate provisions for some identified systems, which can provide insight into the application of prognostics in a graded manner as well as into safety critical systems.

Figure 4 shows the framework for PHM for NPPs. The proposed approach, while utilizing the data and information that is available in the traditional approach, envisages development of prognostic sensor systems to monitor the identified precursor parameters. The data available through the sensors are mapped on the prognostic algorithms to track deviation and therefore provide information on incipient faults. Here, the role of intelligent tools like Support Vector Machine or Bayesian estimation or Sequential Probability Ratio Technique is to predict the prognostic distance such that action can be taken well before the situation results into safety or availability consequences.

The following subsection deals with major aspects of prognostic implementation which are relevant to NPPs.

6.2 Prognostic Approaches

6.2.1 Probabilistic or Reliability-based Approach

This approach is used extensively to predict the life and reliability of components, be they mechanical, electrical, or electronic components. Even though this is considered to be an approximate approach, the advantage of this method is that uncertainty characterization comes naturally. Often the Weibull distribution is utilized extensively. Other distributions, such as exponential and log-normal distribution, are also common as a life prediction model (Yates et al., 2006; Modarres, Kiminskiy and Krivstov, 2010). The weakness of this approach is that the predictions are based on the past performance data of equipment and components. This implies that the prediction does not account for changed component operational and environmental loads. For example, for a given component in a component database, the failure rate estimations are based on an operational environment where the average temperature and relative humidity is 28°C and 65%, respectively. The condition for which the failure rate estimation is required to work is a ground benign environment of 22°C and humidity of 55%. These environmental conditions are bound to affect the failure rate—in this case, reduction in failure rate. Certain external factors such as vibration and seismic shocks adversely affect the life and performance of a component. If these aspects are not factored into the estimates based on historical data, then the estimates tend to be either optimistic or conservative, depending on the severity levels of the component in the database compared to the component for which failure rates are being estimated. If a given component experiences less vibration and seismic shock than a component with a failure rate estimate based on higher vibration and shocks, then the failure rate estimates will not be accurate. Often these types of situations involving application of PHM approach in real-life situations are handled by providing uncertainty bounds.

This approach involves prediction of the mean life of a component along with its upper and lower uncertainty bounds. A wide uncertainty bound indicates that the prediction is based on limited data sets, that reliance on such estimates should be lower, and that these estimates should be used as an indicator. In such situations, precursor-monitoring techniques such as vibration or temperature monitoring of the components represent an effective strategy for prognostics. The Bayesian model features probabilistic estimates that form a priori and has data coming from the precursor monitoring that can be used as evidence for updating the strategy for prediction (Yates and Mosleh, 2006). So, even though the approach is primarily probabilistic in nature, trend monitoring is used to improve the prediction capability.

6.2.2 Physics-of-Failure-based Approach

The physics of failure (PoF) approach deals with the application of first principle models to understand the various failure mechanisms and thereby predict the remaining useful life and reliability of components. In other words, this approach is based on the development and application of scientific models that predict the life of component. In this approach, unlike statistical approaches for reliability estimation, past performance data is not required (White and Bernstein, 2010). The predictions are based on the component characteristics, such as material properties, geometrical attributes, and activation energy for applicable degradation processes for given environmental, operational, and environmental stressors. Accelerated life testing is central to the PoF approach. PoF enables the identification of dominant failure modes and mechanisms, and thereby precursors for monitoring the health of the component. Failure Modes, Effects, and Criticality Analysis (FMECA) form the cornerstone of this approach to identify and prioritize the applicable degradation mechanisms. Identification of precursors is an important part of this approach. The precursors are the parameters that can be monitored using the available sensors (Patil, Das, Pecht, Celaya and Goebel, 2009). Precursor monitoring provides advanced information about the underlying degradation mechanism.

The PoF model can be expressed in general form as:

$$t_{50} = f(x_1, x_2, x_3 \ldots \ldots) \qquad [1]$$

where t_{50} is the median life and x_i is the parameter of the model. The commonly known PoF model for life prediction is the Arrhenius model, which is expressed as follows:

$$t_{50} = A \exp\left[\frac{E_a}{kT}\right] \qquad [2]$$

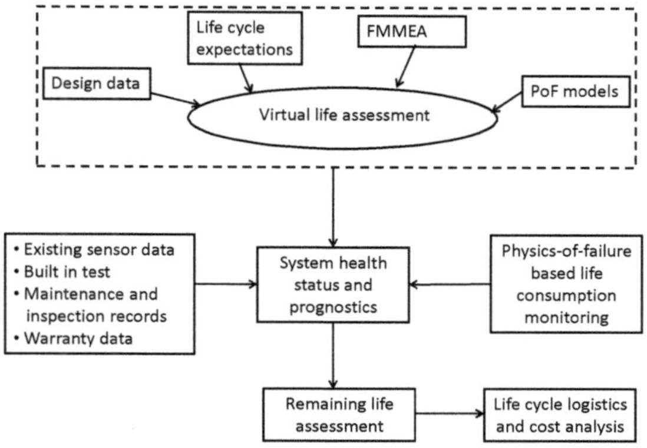

Figure 5. CALCE Physics of Failure Based Approach for Prognostics (Pecht and Gu, 2009).

where A is a process constant, E_a is the activation energy of the process in eV (electron volt), k=Boltzmann's constant= $8.617*10^{-5}$ eV/K, and T is the temperature in Kelvin.

There are many more models available to predict life, such as the Eyring model. The limitation of these models is that they only recognize temperature as an environmental stress. In real life, there are many environmental and operational stresses. These models also fail to account for the geometrical and other material mechanical design features such as material finishes and materials of the mating parts. Overall, the challenge translates into assessment of the accurate prediction of activation energy for a given case.

These limitations have led to further research into developing PoF models that take under consideration the various stresses associated with each component in reliability modeling. This became possible with a basic understanding of the physics of degradation of materials under various stresses. Accordingly, accelerated life testing has become central to PoF modeling. Root cause analysis is performed to understand the degradation mechanisms responsible for failure. Figure 5 shows the framework of PoF approach developed at Center for Advanced Life Cycle Engineering (CALCE), University of Maryland, USA (Pecht, 1996).

This figure shows that information about the component, including physical specifications, geometry, construction materials, operating environment (including temperature, humidity, and vibration), and operational stresses (current, voltage, and electric field) forms the main input for modeling.

One of the notable and significant features of the PoF approach is the development and application of canaries (Dasgupta, Doraisami, Azarian, Osterman, Mathew and Pecht, 2010). Canaries are miniature versions of the subject electronic component, which is designed to fail early. This early failure predicts the impending failure of the subject component. The available literature provides examples of the application of canaries for prognostics (Pecht, 2008). It is very important to note that the word "canary" was coined very recently for incorporating a "weak link" or "weak device" into electronic systems. However, the concept of weak link has been used in mechanical and electrical systems to protect major failure in these systems due to over stresses. In electrical systems the "fuse" protects the electronic or electrical system by cutting of power when electrical stresses reach above the pre-designed levels. Similarly, when mechanical components, such as fuel elements that are incorporated with tension members or pins which are designed to fail, before permanent damage occurs to the reactor structural components or in the fuel itself due to over stressing during fuel handling. This weak link concept, coupled with the present knowledge base and further R&D on prediction of remaining life, provides a promising approach to developing canaries for mechanical and electrical systems.

The PoF approach is particularly suited for assessing electronic component reliability. Even though this approach is in the research and development stage, there are many models available for electronic components. The state of the art in micro-electronic reliability shows that greater advances have been made for reliability modeling of micro-electronic components compared to power electronic components.

6.2.3 Mechanics of Failure (MoF) Approach

The root cause failure analysis of mechanical components and structures may differ from the RCA of electronic components. In mechanical components, the RCA often deals with the macro-level. In a few cases the micro-level of investigations is necessary for understanding failure mechanisms, unlike for electronic components, which require developing models and methods that function at the micro- and at nano-levels. RCA of mechanical components may not always require high precision lab facilities, tools, methods, and software. Often, the stress-strength reliability model with prediction capability and within reasonable uncertainty bands may provide satisfactory insights into failure modes and mechanisms. In fact, the failure modes of mechanical components include failed to open, leakage, failure on demand, and blockage. These failure modes can be verified by visual examination, unlike electronic components, where often the information about failure mode is not directly available or based on failure symptoms.

Another important feature included in MoF is a detailed investigation through corrosion-related modeling. These models and investigations deal with various materials, weld joints, environments, stresses (stress corrosion cracking), finishes, and provisions of protection against corrosion.

The failure surface characterization at the macro or micro level, often part of any RCA for mechanical components, provides reasonable results. Apart from this modeling and

simulation, using the finite element approach at the macro-level can provide satisfactory inputs. Generally, an accelerated test approach does not form part of the RCA of mechanical components, unlike electronic components where accelerated life testing forms the bottom line. Even though mechanical components can also be modeled using the PoF approach, the nature of prediction and level of treatment required to model mechanical components and structures require the problem to be handled at the macro-level rather than the micro-level. Hence, this paper proposes a collection of tools and methods through an approach called mechanics of failure (MoF) to predict the life of mechanical and structural components. Here, the stress-strain relationship forms the fundamental approach to reliability and life prediction.

Even though this approach is best suited to tackle one of the main bottlenecks of risk-based approach, namely, assessment of the safety margin, the data and the models available so far often form a limitation to predict reasonably accurate safety margins. Accelerated testing methods, probabilistic fracture mechanics approach, damage mechanics, strength of material methods, finite element analysis, and failure analysis methods form the major elements of the MoF approach. Reliability methods are used as part of the MoF approach for making statistical estimates of life. The major degradation mechanisms that are evaluated in this approach include mechanical wear, creep, corrosion, and catastrophic failure. As with the PoF approach, MoF also utilizes root cause failure analysis models for understanding underlying failure or degradation mechanisms. The work performed by Mathew et al. provides a structural analysis of a structural board for NASA and appears to bring out, in a way, the essence of the MoF approach (Mathew, Das, Osterman, and Pecht, 2006). There are many studies in the literature where the objective is to base the prognostics on two major degradation mechanisms—temperature and vibration (Pecht, 2010). Of course, when the application is designed for nuclear core components, irradiation-induced degradation often becomes the leading parameter (IAEA, 1998).

6.2.4 Symptom or Data-based Approach

Generally, this approach is referred as the data-driven approach. This section deals with a data-driven approach, except that here a distinction between various data forms an input for prognostics. For example, a trend monitoring of operational and environmental parameters through on–line instrumentation may provide information about some precursor. A pattern comprising the status of a finite set of alarms as "registered as 1" and "cleared as 0" is another representation of data. A probabilistic distribution of time to failure based of individual components provides time to failure estimates of the systems being monitored. Input can be in the form of linguistic variables in place of a numerical value. All these require different approaches.

The term "symptom-based approach" is used in this paper to extend the context of input data and information used in prediction, particularly for nuclear plant applications. As mentioned above, often information is not available in the form of a numerical value or in the form of binary values (0/1 or yes/no). Instead the information about the model parameters comes from experts in linguistic expressions. This information is not suitable for use as input; however, the information cannot be ignored, as it provides much stronger input for prediction or estimation of remaining useful life. In such instances, treating expert opening, which can be considered imprecise information, using fuzzy algorithms can provide one with improved assessment of imprecise parameters (Chen and Vachtsevanos, 2012).

Second, the reason to have provision for some information is that establishing a pattern is important, as often, instead of a single parameter, a pattern can provide more data and information. For example, a comparative value of three parallel components seeing the same operational and environmental stresses may form a pattern, which may provide an effective mechanism to assess the health of the component and thereby provide an effective input for predicting the remaining life. The only issue is that even this information could be expressed in terms of linguistic variables and will require the fuzzy approach to address the challenge.

This background is an obvious reason to formulate the data and information in two ways: trend monitoring using precursor symptoms and a pattern-driven knowledge-based approach.

6.2.4.1 Trend Monitoring

Trend monitoring is a natural extension of the condition-based approach to diagnostics. Often pump bearing temperature, vibration reading, or a pump shaft that has run out of measurements forms part of a condition-monitoring program in nuclear plants. An expert can predict the time when a piece of equipment will need to be shutdown. This practice is common in industrial environments in general and nuclear plants in particular. In these cases, pump bearing vibration, temperature, and shaft run-out act as precursors for the prediction.

To extend this approach to a prognostic regime, it is required that the deviations be tracked or monitored in online mode, that the failure criteria and associated uncertainty band be assessed for the component in question, that the future operational and environmental loading be used to assess the remaining useful life, that the prognostic distance, which can come from the maintenance logs, be assessed, that the uncertainty in RUL estimates be assessed; and that the degradation rate and alarm be predicted online as soon as the prediction upper estimates overlap with the lower bound of the failure criteria.

There are many examples of models and methods that have been developed for prognostics and health assessment of check-valves or loose part monitoring in nuclear plants. This approach is particularly useful when the degradation profile is well understood. This means that this approach is more applicable to micro-electronic components where adequate accumulated operational experience on degradation trend is available. The availability of a PoF approach to the modeling of micro-electronic systems is testimony to this observation.

However, when it comes to power electronic components, one can only claim that work on the application of a PoF model for these systems has been initiated (Yin, Hua, Mussalam, Baily and Johnson, 2010; Patil, Das and Pecht, 2012) but still is not as developed as micro-electronic systems. In the absence of the proper understanding of the degradation process, the current strategy to overcome this limitation is to uprate the system by providing the extra margins in design.

6.2.4.2 Pattern-driven Knowledge-based Approach

This approach does not require description of the system or component through basic models, but only requires patterns comprising component and system specific historical data and information. It provides an efficient and effective mechanism where the input/output relation cannot be defined through well-defined scientific models. However, it establishes that there is a one-to-one relationship between a set of input patterns and corresponding states in a system. One example is the alarm/trip pattern in a reactor, which can be associated with a discrete reactor state. When number of patterns or vectors each comprising a set of alarms uniquely define a given reactor state, then this approach can effectively be used for reactor status and condition monitoring. The advantage of this approach is that it can operate successfully even with missing primary data to form a precursor for predictions. As it operates on clusters of data and often derives data it can use to predict with reasonable assurance. The applications include nuclear plant transient identification, prediction reactivity, and health monitoring in rotating machines. This approach has often been implemented using artificial neural networks (ANNs) (Varde, Verma and Sankar, 1998; Lee et. al., 2005), neuro fuzzy systems (Chen, Zhang, Vachtsevanos and Orchard, 2011), support vector machines (Abe, 2010), or sequential probability ratio techniques (Coble et al., 2010). When ANN tools are used, the approach involves training the ANN with various patterns, including healthy patterns and various failure patterns for specific components.

Another example is application of ANN for health prediction for check valves (Lee, Lee and Kim, 2005), (Uhrig, 1994). The ANN is trained with historical data on the failure of the check valve involving the hinge pin, dish, stopper pin, or dash pot. These patterns, along with healthy patterns, are used to train the ANN. Validation and verification of the algorithm is carried out by testing the ANN response for new and existing patterns, for which it has been trained, plus unlearned patterns. The recall tests are often carried out with additional patterns having noise, missing data, or fuzzy data to ensure that the prognostic model is robust and that repeatability is high. During the course of prognosis, if the ANN algorithm comes across a new pattern that was not there in the database, there is a provision to train the ANN for this new condition. This new pattern then becomes part of a pattern-knowledge-based library for prognostics.

It may be noted that the approaches listed above fall in the category of intelligent methods. The objective is to extract the features specific to a given input pattern. Here, the main issue is to determine which approach should be used for arriving at a given solution. Often, this decision comes from assessing the nature and complexity of the level of details that are expected in the solution space. This often requires performance evaluation of the approaches under consideration (Varde et. al, 1998).

6.2.5 Integrated Approach.

The prognostics approach followed for electronic components often uses what is called a fusion approach to enhance the prediction capability of the prognostics approach. General experience has been that often one approach may not be adequate to provide the desired results. Hence, the trend-driven approach is integrated with the PoF approach. While the PoF approach prides fundamental requirements for prognostic models, the database approach complements the model with a knowledge base that has already been developed for various failure modes.

An integrated approach is also beneficial where the available data is inadequate to implement prognostics. To improve the prediction accuracy and precision, it is often necessary to use Bayesian updating to incorporate new data for prediction (Modarres et al., 2010). Hence, the probabilistic approach is used in conjunction with online precursor trends to update estimates with new data available from online sensors. While the trend monitoring identifies the deviation from the normal operation of equipment, the probabilistic model with uncertainty bands will provide an estimate of the prognostic distance—a performance metric crucial for fixing the deficiency either through repair or replacement. The prognostic distance also prompts the plant manager to plan the action in advance such that plant availability and safety can be optimized.

6.3 Material Degradation and PHM Requirements

Nuclear power plants include PWR, BWR and other designs such as Canadian deuterium reactor (Candu), pressurized heavy water reactors (PHWR), and gas-cooled reactors. The accumulated operating experience works out to be 10,750 reactor years, considering an average operating experience of 25 years. Logs of failure history of components provide

indicators for degradation trends. Prognostic applications require research and development to fuse the historical data with the available PoF models, considering intrinsic and extrinsic parameters, to gain improved understanding of the degradation of SSCs. It may be noted that non-destructive evaluation or testing (NDE or NDT) forms a major component of the surveillance of SSCs in NPPs (Baskaran, 2000). With respect to the aging or degradation of SSCs, degradation is a slow and gradual process and the prognostics used to track trends exists only after a period of 30 years (Bond, 2008a). This means that the pre-service inspection (PSI) data collected during the plant licensing phase forms a template or reference for future trend monitoring.

However, often the information may not be adequate to provide support of the estimation of remaining useful life or to determine the failure criteria for a given material application. This is why there is an overwhelming desire to have a proactive approach to the management of material degradation in nuclear plants in general and aged plants in particular (Bond, 2008a; Bond, 2008b). The incentive for the operating organization is to support the case for life extension while for regulators it provides flexibility for oversight and monitoring to generate feedback. Hence, monitoring as part of the implementation of PHM strategy backed up by degradation models forms a vital element for remaining useful life prediction (Meyer, Ramuhalli and Bond, 2011).

For a prognostic strategy to be effective, it is important to have the reference signatures of the systems during the initial stages of operation. This makes it prudent that all the condition monitoring and surveillance applications, like leak-before-break, coolant channel health monitoring, installation of coupons (to assess corrosion in strategic location for in-core or out-of-core components), in-service-inspection, bearing signatures, and cable insulation strength be seen as prognostic applications. As the life of components depends on lifetime loading and variation in environmental, electrical and mechanical stresses, (this includes the effect of external events and new combined phenomenon), it is important that PoF and MoF models account for degradation history to predict life.

6.4 Prediction and Learning Machines and Tools

There are a host of approaches for prediction, including probabilistic and statistical approaches. Examples of probabilistic methods include regression modeling, Bayesian updating (Guan et al., 2011; Modarres et al., 2010), principal component analysis and sequential probability ratio tests (Coble, 2010). The intelligent methods or machine learning approaches form an important element in prognostics. The common approaches employed for prediction and machine learning are artificial neural networks (ANNs) (Varde et al., 1998), neuro-fuzzy models (Chen et al., 2010; Chen et al. 2011), Kalman filters

(Heimes, 2008), particle filters (Chen, Vachtsevanos, and Orchard, 2010), and support vector machines (Abe, 2010). The selection and application of a given approach is based on the nature of the predictions to be made. For example, ANN tools are used when the prediction is based on symptoms and not an actual model of the system. The Bayesian approach is used where the prior knowledge predictions are based on new evidence or data. Probabilistic approaches are the traditional methods for estimating time to failure as an indicator of remaining life. Support vector machines are used where predictions are to be based on the clustering of data and information to form patterns.

6.5 Limitation of Prognostic Methods

Even though prognostics has evolved into a relatively new paradigm with applications in areas such as space, aircraft, and structural engineering, the development and deployment of prognostics in NPPs is very limited (Shafto et al.,). There are certain issues that need to be addressed through research and development efforts.

Major challenges to the implementation of prognostics include sensors and associated networks, PoF and damage models and failure criteria, uncertainty characterization, and organizational frameworks. The availability of sensors in general and the development of an integrated sensor network can be considered one bottleneck in the implementation of prognostics. This is particularly true for electronic components, as this application requires miniaturization of the sensors such that newly developed sensors and networks can fulfill the requirements of an application. Keeping in view the enhanced performance of prognostics for future applications, wireless sensor networks (WSNs), along with utilization of miniaturized sensors such as Pt-100 for online temperature measurement, provide with an effective technique for the implementation of prognostics for electronic components (Puccinelli and Haenggi, 2005), (Lin, Wang, and Sun, 2004).

The designers of newly built plants have to take a proactive approach for making prognostics provisions. This requires focused efforts on preparing design specifications based on safety and availability studies that identify not only components and processes that require PHM, but also selecting a PHM approach depending on failure mechanisms. For instance, if prognostic provisions are required for certain in-core components, suitable provisions should be made right before the start of construction activity. For existing plants, plant constraints will dictate the level of prognostics to be implemented. However, when life extension is being explored for the new plants, implementation of prognostics tools and methods can provide valuable insights into tracking aging mechanisms as well as help in assessing performance of systems in on-line mode or at periodic intervals. Keeping in mind advances in wireless sensor network (WSN) applications, the pros-and-cons of this technology should be evaluated. On the one

hand, while there is immense potential for WSN technology, there are some limitations, which include, a) lower speed, b) requirement of power supply to the node, c) more complex to configure and d) the performance of the node is easily affected by surroundings like walls, microwaves, large distances due to signal attenuation, etc. (Bhattacharya, Kim and Pal, 2010).

In the nuclear industry the principles of defense in depth ensures the implementation of redundant and diverse electronic channels; however, the common cause failure (CCFs) aspects require special attention. The effect of any degradation or failure mode, induced by material, environmental, operational parameters needs to be analyzed, particularly for assessing its CCF impact as part of PHM implementation (IAEA, 2009). For developing the PoF model for electronic protection channels, the potential failure due to whisker growth, electromigration induced shorting of parallel metallization, coupling of the redundant path due to field effects and solder joint failure requires special CCF considerations.

Similarly, for developing degradation models for in-core structural components as part of PHM implementation requires not only the monitoring provisions that include special sensors but also considerations and development of irradiation induced degradation and growth models. The prediction accuracy of these models will require assessment of change in material property in dynamic manner with the fluence it has seen in the reactor core (IAEA, 1999; Dharmaraju et al., 2008). When implemented, these models are expected to provide input for a risk-based approach (Samal, 2010).

Often if the PoF models/damage models are available it is challenging to define the failure criteria and the uncertainty associated with these definitions. The lack of knowledge related to failure criteria is often addressed by conservative assumptions. Here, the role of prognostics becomes crucial, as the online signal can be used with the available data and models to characterize the incipient failures.

There are two types of uncertainties that need to be addressed in prognostics: aleatory and epistemic. Aleatory uncertainty, which is inherent in nature and cannot be reduced, arises from data and models. Epistemic uncertainty is uncertainty, which is reduced by acquiring additional knowledge or data. The integrated approach is a typical example of reducing epistemic uncertainty. Reducing uncertainty in PHM becomes more important from the point of estimating prognostic distance. At a higher level, it affects the accuracy of the assessment of the safety margin as part of risk-based applications. Other approaches to model or reduce uncertainty involve updating the prior data with new evidence using well-known techniques such as Bayesian updating, Kalman filtering, constrained optimization, and particle filtering.

In spite of the above developments, the accuracy of uncertainty assessment is a lingering issue. Other non-parametric methods that are expected to reduce subjectivity in uncertainty assessment are being developed. One method is the imprecise probability based approach. However, there are limited applications of this approach. Further R&D in this area may provide a new approach to uncertainty modeling and analysis.

PHM is a resource-intensive application. Hence, organizational will to implement and operate a PHM program is a pre-requisite. Whether it is for routine health management of components in support of surveillance or life extension studies for new plants, the involvement of not only implementation-level staff but also plant management is an important factor in the success of the PHM approach (Pecht, 2010).

The availability of a PoF or damage model is one of the major challenges to the initiation and implementation of a PHM program in a nuclear plant. Even though limited application up to condition monitoring has found wider application in the nuclear sector, the real benefits or full potential of prognostics can be realized only after the damage model for mechanical and structural engineering components and PoF models for power and micro-electronics and electrical components.

5. PERFORMANCE CRITERIA

Performance criteria or metrics determine the adequacy of a prognostic approach for a given application (Saxena et al., 2010). Extensive work has been reported to define the appropriate performance metrics for a given application to improve the performance of prognostics and health management and condition monitoring approaches (Wheeler et al., 2010; Saxena et al., 2010; Pecht, 2009; Feldman et al., 2010; Coble and Hines, 2008). As the literature shows, there are three major performance indicators to determine the efficiency and effectiveness of PHM applications: prognostic distance (PD), accuracy, and precision.

5.1. Prognostic Distance

The time between the predicted time of incipient failure and actual component failure is called the prognostic distance. This definition has been derived from the concept of prognostic distance used in canaries (Wang, Luo, and Pecht, 2011). This indicator can be better understood by the following assumptions and observations: 1) assume that the time t_{af} is the actual time to failure of a component involving a particular failure mechanism, 2) t_m is the time for completion of the recovery of a component, assuming that the time for actual failure is known, 3) the time when the prognostic model detects and confirms the degradation trend as positive is t_d, 4) the prognostic algorithm has the built-in capacity to predict uncertainty in the remaining life estimate and degradation parameter assessment, and 5) the

t_0 = 0; Time when component was put in service after test or maintenance

t_{af}= Time of actual failure

t_m= Epoch of time when maintenance / recovery should complete

t_d= Prediction of Failure by Prognostic algorithm

t_{ep}= Early prediction of deviation

t_m - t_d= Time required for reconfiguration / recovery action = A

t_{mg} - t_{af}= Time available for repair / recovery /mitigation (referred as plant coping time) = B

t_d - t_{mg}= Time available for repair / recovery /mitigation (referred as plant coping time)=C

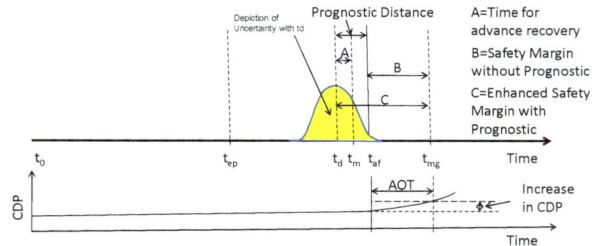

Legends: CDP: Core Damage Probability
AOT: Allowable outage time

Figure 6. Depiction of features of prognostics and performance criteria.

success of metrics is determined by how far in advance it predicts the deviation such that an adequate time window is available for the replacement or recovery action such that availability and safety functions are ensured. However, the prediction should not be so early that it results in a loss of component life or premature replacement. Fig. 6 depicts the condition for optimum prognostics.

7.2 Accuracy: Accuracy means the correctness of the remaining life estimates. As can be seen in Figure 6, the correctness of the prediction of time determines the accuracy of prediction.

7.3 Precision:

Precision accounts for the uncertainty estimates in remaining life prediction. The width of the uncertainty band determines the precision of the estimates. A shorter band has higher precision, and a wider band has lower precision.

Other parameters that are of interest to risk-based applications include assessment of the safety margin for the case or scenario being evolved. Figure 6 also shows the increased safety margin made available by the prognostic algorithm. The increase in core damage probability (CDP) is depicted by the lower time vs CDP plot. The plant technical specification defines the allowable outage time. It can be seen that by keeping a safety related system or component, the core damage probability increases while the process of prognostics is dynamic in nature, the efficiency and effectiveness of this process, from the risk evaluation point of view, is determined by how effectively the process addresses performance trending and follow-up activities. This increase is linear with time. This aspect extends the role of a prognostic algorithm to monitoring and comparing the performance of the subject case or component to ensure that it meets the performance criteria set or recommended by the regulator. The algorithm then produces documentary evidence, providing the estimates of assessment of safety margin, characterization of uncertainty, critical parameter trends, and projected life of the new modifications. These indicators are of particular interest to risk-based applications.

6. CONCLUSIONS AND RECOMMENDATIONS

There is increasing use of condition monitoring in support of operation and maintenance of nuclear plants. The diagnostic and prognostic approach can be used as part of a risk-based approach. A risk-based approach can support the prognostics program. Looking at the publications in the areas of mechanical and structural engineering, it can be argued that a prognostics framework for nuclear plants can be established by adopting the models and methods developed for space, aircraft, and civil engineering systems for core components where radiation-induced degradation may not play much of a role in dictating the remaining useful life. For core components, a limited knowledge base is available that can be utilized with certain uncertainty bounds.

We have proposed a new paradigm called the mechanics of failure as part of prognostics implementation. The MoF approach to a large extent operates in the manner of a PoF approach; the only difference is that most of the times MoF deals with macro- or micro-level analysis tools and methods. It has been argued out that although PoF is more suitable for the modeling and analysis of electronic and power electronic components, MoF works for structural systems and components in nuclear systems, such as pressure vessels, coolant channels, pumps, valves, pipes, and heat exchangers.

Prognostics can be applied to new plants by making the complete monitoring and surveillance and maintenance management process more effective through the prediction of fault and degradation trends, such that adequate time is available to have recovery and repair action. This aspect is important as it works for both safety and availability improvement. For existing or older plants with constraints imposed by design, layout, or operational limitations, this approach is expected to be very effective for life extension studies that are carried out as part of aging studies and performance monitoring after the changes and modifications have been incorporated. As can be seen, all of these gains go further towards consolidating the risk-based approach.

Advances in any field and their application to real-life situations are normally judged by the availability of codes and standards. Even though there are many standards and codes for surveillance and condition monitoring, there are hardly any standards on prognostics and health management. In this direction, the development of the first IEEE prognostics and health management standard for

electronics is at an advanced stage and appears to be undergoing review (IEEE). The availability of this standard will mark a significant step: it will channelize the knowledge base available in advanced labs for system applications to industry. As far as the nuclear industry is concerned, this will be a clear incentive to develop prognostics for electronics in reactor controls and protection.

Based on the review of the status of existing surveillance and monitoring program and the potential role that prognostics can play as part of the IRBE in NPPs, we make the following recommendations:

A prognostics approach brings in the element of dynamics into the existing risk-based approach. Hence there is a strong argument in favor of initiating a prognostics-based health management program in NPPs. The current knowledge of prognostics is such that extensive research and development is required, particularly for power electronic systems and electrical systems, such that accurate prediction of remaining useful life can be developed.

The development of PoF and MoF models require elaborate life test set-ups and material characterization facilities. Apart from this, the study of irradiation-induced degradation requires research reactor test facilities. The available resources can be networked in a coordinated manner to support this development work. There should be provision in the operating reactor organization to communicate data and insights on failure to a prognostics laboratory on the one hand while providing prognostic solutions for real-time issues on the other.

Even though the prognostic approach for estimating the life and reliability of the components in new and old plants remains similar, the emphasis in new plants is to develop a host of prognostic performance metrics for the identified components while for old plant prognostics must support inspection, testing, and condition monitoring. Development efforts should adopt the prognostic systems that have been developed in other fields, such as navigation, aircraft, space, and infrastructure systems, so that the program is more effective in terms of deliverables.

Nuclear research labs generally have a reasonable infrastructure for developing prognostic sensors and associated systems. Hence, early identification of sensor requirements is important for the success of prognostic programs.

Work should start on the development of codes and standards for prognostics and health management for nuclear components and systems.

Keeping in mind the benefits that can be realized through the implementation of a risk-based prognostic program, this paper argues a case for setting up centers of excellence to facilitate research and development on prognostics for engineering systems in complex systems such as nuclear power plant. This is required as the prognostic and health management approach has the potential to benefit existing plants entering the aging zone and new plants where a target life of more than 90 years can be met with online prognostics that enable degradation monitoring of critical systems and components.

REFERENCES

Abe Shigeo, (2010). *Support vector machines for pattern classification - advances in computer vision and pattern recognition.* Second Edition, Springer.

Andonov, A., Apostolov, K., Kostov, M., A. Iliev, Stefanov, D., Varbanov, G., (2011). Structural health monitoring of VVER-1000 containment structure. Transactions, *SMiRT 21,* New Delhi, India

Baskaran S., (2000). Role of NDE in residual life assessment of power plant components. *NDT.net,* Vol. 5 No. 07.

Bhambra J.K., Nayagam S., Jennions I., (2011). Electronic prognostics and health management of aircraft avionics using digital power convertors. 2011 *Annual Conference of the Prognostics and Health Management Society.*

Bhattacharya, D., Kim T., and Pal S., (2010). A comparative study of wireless sensor networks and their routing protocols. *Sensors, 10, 10506-10523; doi:10.3390/s101210506,* ISSN 1424-8220.

Bond L.J., Doctor. S.R. and Taylor, T.T., (2008a). Proactive management of materials degradation – a review of principles and programs. *PNNL-17779, Pacific Northwest National Laboratory, U.S. Department of Energy, Richland, Washington.*

Bond L.J., Taylor, T.T., Doctor S.R., Hull Amy B., and Malik S. N., (2008b). Proactive management of materials degradation for nuclear power plant systems. *2008 International Conference on Prognostics and Health Management.*

Chatterjee, K, and Nodarres, M., (2012). A probabilistic physics-of-failure approach to prediction of steam generator tube rupture frequency. *Nuclear Engineering and Science,* Vol. 170, Number 2, pp. 136-150.

Chen Chaochao, Brown D., Sconyers Chris, Zhang Bin, Vachtsevanos George, Orchard M.E. (2012). An integrated architecture for fault diagnosis and failure prognosis of complex engineering systems. *Expert Systems with Applications 39 9031-9040.*

Chen Chaochao, Pech Michael, (2012). Prognostics of lithium-ion batteries using model-based and data-driven methods. *Prognostics and System Health Management Conference* (PHM-2012 Beijing).

Chen Chaochao, Vachtsevanos George (2012). Bearing condition prediction considering uncertainty: an interval type-2 fuzzy neural network approach.

Robotics and Computer-Integrated Manufacturing 28 509–516.

Chen Chaochao, Vachtsevanos George, and Orchard Marcos E. (2010). Machine remaining useful life prediction based on adaptive neuro-fuzzy and high-order particle filtering. *Annual Conference of the Prognostics and Health Management Society.*

Chen Chaochao, Zhang Bin, Vachtsevanos George, and Orchard Marcos (2011). "Machine condition prediction based on adaptive neuro–fuzzy and high-order particle filtering. *IEEE Transactions on Industrial Electronics,* Vol. 58, No. 9, 4353.

Coble J.B. and Hines J.H., (2008). Prognostic algorithm categorization with PHM challenge application. *2008 Int. Conf. on Prognostics and Health Management.*

Coble, J., and J.W. Hines, (2010). Application of failure prognostics to the IRIS plant., *Seventh American Nuclear Society International Topical Meeting on Nuclear Plant Instrumentation, Control, and Human-Machine Interface Technologies NPIC and HMIT 2010,* Las Vegas, NV, Nov. 7-11.

Coble, Jamie, Humberstone Matt, and Hines J. Wes, (2010). Adaptive Monitoring, Fault Detection and Diagnostics, and Prognostics System for the IRIS Nuclear Plant. *Annual Conference of the Prognostics and Health Management Society,* Portland, Oregon, USA.

Coble, JB, Ramuhalli, P, Bond, LJ, Hines, JW, and Upadhyay, BR, (July 2012). Prognostics and health management in nuclear power plants: a review of technologies and applications. *PNNL-21515,* Contract DE-AC05-76RL01830, Pacific Northwest National laboratory, Richland, Washington.

Coppe A., Haftka R.T., Kim Nam-Ho, and Bes C. (2008). "A statistical model for estimating probability of crack detection. *2008 Int. Conf. on Prognostics and Health Management.*

Dasgupta Abhijit, Doraiswami Ravi, Azarian Michael, Michael Osterman, Sony Mathew, and Michael Pecht, (2010). The use of "canaries" for adaptive health management of electronic systems. ADAPTIVE 2010 : *The Second International Conference on Adaptive and Self-Adaptive Systems and Applications,* IARIA, 2010 ISBN: 978-1-61208-109-0.

Dharmaraju, N. and Rama Rao, A. (2008). Review article: Dynamic Analysis of Coolant Channel and Its Internals of Indian 540MWe PHWR Reactor. *Hindawi Publishing Corporation - Science and Technology of Nuclear Installations,* Volume 2008, Article ID 764301, 7 pages, doi:10.1155/2008/764301.

Gregor, F and Chokie, Alan (2006). Ageing management and life extension in the US nuclear power industry. *CGI Report 06:23, Prepared for the Petroleum Safety Authority Norway, Chokie Group International Inc, USA.*

Feldman Alexander, et. al., (2010). Empirical Evaluation of Diagnostic Algorithm Performance Using Generic

Framework. *International Journal of Prognostics and Health Management*", 002, (ISSN 2153-2648).

Gu Jie, Vichare Nikhil, Tracy T. and Pecht, M. Prognostics implementation methods for electronics. http://www.prognostics.umd.edu/calcepapers/07_Jie_Pr ognosticsImplemntationmethodElectronics_RAMS.pdf.

Guan X., Liu Y., Jha R., Saxena A., Celaya J, Geobel K., (2011). Comparison of two probabilistic fatigue damage assessment approaches using prognostic performance metrics. *International Journal of Prognostics and Health Management*, 005, (ISSN 2153-2648).

Hashemian, H.M. On-line Monitoring and Calibration Techniques in Nuclear Power Plants. *IAEA-CN-164-7S05*, IAEA, Vienna. http://www-pub.iaea.org/MTCD/publications/PDF/P1500_CD_Web /htm/pdf/topic7/7S05_H.Hashemian.pdf.

Heimes F.O., (2008). Recurrent neural networks for remaining useful life estimation. *2008 Int. Conf. on Prognostics and Health Management*.

Heng Aiwina, Zhang Sheng, Tan, Andi CC, and Mathew Joseph, (2009). Review – Rotating machinery prognostics: state of the art, challenges and opportunities. *Mechanical Systems and Signal Processing*, 23, 724-739. Doi:10.1016/j.ymssp.2008.06.09.
http: //standards.ieee.org/develop/wg/PHM.html.

Hyers, R.W., McGowan, J.G., Sullivan, K.L, Manwell, J.F., and Syrett, B.C., (2006). Condition monitoring and prognosis of utility scale wind turbines, *Energy Materials*, Vol. 1, No. 3, 187.

IEEE Standard Association, Draft of "*P-1856 – Standard framework for prognostics and health management of electronic systems*, IEEE Reliability Society,

International Atomic Energy Agency (1993). Risk based optimization of technical specifications for operation of nuclear power plant. *IAEA-TECDOC-729*, IAEA, Vienna.

International Atomic Energy Agency (1995). Management of research reactor ageing . *IAEA-TECDOC-792*, IAEA, Vienna.

International Atomic Energy Agency (1999). Assessment and management of ageing of major nuclear power plant components important to safety : CANDU pressure tubes. *IAEA-TECDOC-1037*, IAEA, Vienna.

International Atomic Energy Agency (1999). Living probabilistic safety assessment (LPSA). *IAEA-TECDOC-1106*, IAEA, Vienna.

International Atomic Energy Agency (2009a). Protecting against common cause failures in digital I&C systems of nuclear power plants, *IAEA-Nuclear Energy Series* No NP-T-1.5, IAEA, Vienna.

International Atomic Energy Agency (2009b). Proactive management of ageing for nuclear power plants, *IAEA-Safety Report Series No. 62*, IAEA, Vienna.

International Atomic Energy Agency, (1998). "Assessment and management of ageing of major nuclear power plant components important to safety: CANDU pressure tubes. *IAEA-TECDOC-1037*, IAEA, Vienna.

International Atomic Energy Agency (2010). Risk informed in-service inspection of piping systems of nuclear power plants: process, status, issues and development. *IAEA nuclear energy series no. NP-T- IAEA*, Vienna.

Kadak, Andrew C., Matsuo,Toshihiro, (2007). "The nuclear industry's transition to risk-informed regulation and operation in the United States. *Reliability Engineering and System Safety* 92 609–618.

Kalgren P.W, et. al., (2010). Application of prognostic health management in digital electronic systems. *International Journal of Prognostics and Health Management*, 002, (ISSN 2153-2648). 1-4222-0525-4/07 2007, IEEE Paper 1326 Version 3.

Klein R., Rudyk E., Masad E., and Issacharoff M., (2011). Model based approach for identification of gears and bearings failure modes. *International Journal of Prognostics and Health Management*", 009, (ISSN 2153-2648).

Lee Min-Rae, Lee Joon-Hyun and Kim Jung-Teak, (2005). Condition monitoring of a nuclear power plant check valve based on acoustic emission and a neural network. Jr. of Pressure vessel Tech., issue 3, Volume 127.

Lee, WW, Nguyen, LT, and Selvaduray, (2000). Solder joint fatigue models: review and applicability to chip scale packages. *Microelectronics reliability*, 40, 231-244.

Lin R., Wang Z., and Sun Y., (2004). "Wireless sensor network solutions for real-time monitoring of nuclear power plants., *Proc. 5th World Congress on Intelligent Controls and Automation*, Vol 4, pp. 3663 – 3667.

Mathew Sony, Das Diganta, Osterman Michael, Michael Pech, (2006). Prognostics assessment of aluminum support structure on a printed circuit board. *ASME Journal of Electronic Packaging*, , Vol. 128 / 339.

Meyer, Ryan, Ramuhalli Pradeep, Bond L.J., (2011). Developing effective online monitoring technologies to manager service degradation of nuclear power plants. *IEEE PHM Conference*, Denver, Colorado, USA, 20-23, , Denver.

Mishra S., Ganesan S., Pecht, M, and Xie Jingsong,(2004). Life consumption monitoring for electronics prognostics. *2004 IEEE Aerospace Conference Proceedings*.

Modarres M., Kaminskiy M. and Valiliy Kristov, (2010) *"Reliability engineering and risk analysis – a practical guide"* (Second Edition), CRC Press.

Nuclear Energy Agency, (2005). CSNI Technical Opinion paper No *7 on Living PSA* and its Use in the Nuclear Safety Decision-making Process /No 8 on Development and Use of Risk Monitors at Nuclear Power Plants. *NEA 4411,* Organization for Economic Co-Operation and Development, Paris.

Patil Nishad, Das Diganta, and Pecht Michael (2012). A prognostic approach for non-punch through and field stop igbts. *Microelectronics Reliability*.

Patil Nishad, Das, Diganta and Pecht Michael, Jose Celaya and Goebel Kai, (2009). Precursor parameter identification for insulated gate bipolar transistor (IGBT) prognostics. *IEEE Transactions on Reliability*, Vol. 58, No. 2, pp. 271-276.

Pecht, M.G., (2008) *"Prognostics and health management of electronics"*, Wiley and Sons.

Pecht, Michael and Dasgupta, Abhijit, (1996). Physics-of-failure: an approach to reliable product development, *IEEE - 95* IRW FINAL REPORT.

Pecht, M. and Gu, Jie, (2009). Physics-of-failure-based prognostics for electronic products. *Transaction of the Institute of Measurements and Control, 31, 309-322.*

Pecht, Michael, (2009). Prognostics and Health Management. *Encyclopedia of Structural Health Monitoring,* Ed. Boller Christian, Chang Fu-Kuo and Fujino Y.; John Wiley and Sons, Ltd.

Pecht, Michael G. (2010) *"Prognostics and health monitoring in complex engineering systems: methods and applications"*, IEICE Fundamental Reviews Vol. 3 No. 4.

Puccinelli D., Haenggi M., (2005). Wireless sensor networks : application and challenges of ubiquitous sensing. Feature Article, *IEEE Circuits and Systems*, Third Edition.

Samul, MK, Dutta, BK, and Kushwaha, (2010). A probabilistic approach to evaluate creep and fatigue damage in critical components. *Transaction of Indian Institute of Metals*, Vol. 63, Issues 2-3, pp. 595-600.

Saxena A., et. al. (2010). Metrics for Offline Evaluation of Prognostic Performance. *International Journal of Prognostics and Health Management"*, 001, (ISSN 2153-2648).

Saxena A., Simon D., (2008). Damage propagation modeling for aircraft engine run-to-failure simulation. *2008 Int. Conf. on Prognostics and Health Management.*

Shafto M., "Prognostics Performance Evaluation", National Aeronautics and Space Administration web-site: http://ti.arc.nasa.gov/tech/dash/pcoe/prognostics-performance-evaluation/metrics/algorithm-performance/.

Sinha, R.K., Sinha, S.K., Madhusoodhan, K., (2008). Fitness for service assessment of coolant channels of Indian PHWRs. *Journal of Nuclear Materials*, Vol. 383, Issue 1-2, pages 14-21.

Tantawy A., Koutsoukos X., and Biswas G., (2008). Aircraft ac generators: hybrid system modeling and simulation. *2008 Int. Conf. on Prognostics and Health Management.*

TSU-MU KAO, (2007) *"Risk-Informed Regulation and Applications in Taiwan"* International Journal of

Performability Engineering Volume 3, Number 1, - Paper 5 - pp. 47 – 59.

U.S. NUCLEAR REGULATORY COMMISSION, (2002). Regulatory guide 1.182 - assessing and managing risk before maintenance activities at nuclear power plants. *Office of nuclear regulatory research*, Washington.

Uhrig, R.E., (1994). Application of artificial neural network in industrial technology. *Proc. of the 1994 IEEE International Industrial Technology Conference.*

Varde, P.V. Sankar S. and Verma, A.K., (1998). An operator support system for research reactor operations and fault diagnosis through a connectionist framework and PSA based knowledge based systems. *Reliability Engineering and System Safety*, 60, 53 – 69.

Varghese, Joy P., Verma, V.S., Rajput, C.D., Ramamurthy, K., (2010). *"En-Masse Coolant Channel Replacement in Indian PHWR.* Energy Procedia *00 000-000.*

Vichare, N.M. and Pech M.G.,(2006). Prognostics and Health Management of Electronics. *IEEE transactions on components and packaging technologies*, V:29, 1.

Wang, W., Luo S., and Pecht, M.G. Economic design of the mean prognostic distance for canary-equipped electronic systems. *Microelectronics Reliability, 52 (2012) 1086-1091.*

Wheeler K.R., Kurtpglu T., and Poll S.D., (2010). A Survey of Health Management User Objectives in Aerospace Systems Related to Diagnostic and Prognostic Metrics. *International Journal of Prognostics and Health Management"*, 003, (ISSN 2153-2648).

White M., Bernstein, (2010). Microelectronics reliability: physics-of-failure based modeling and life time evaluation. *NASA Electronic Parts and Packaging (NEPP) Program Office of Safety and Mission Assurance*, NASA EBS: 939904.01.11.10, Jet Propulsion Lab, 4800 Oak Grove Drive, Pasadena, CA.

Wood S.M., Goodman L.D., (2006). Return-on-investment (roi) for electronic prognostics in high reliability telecom applications. *1-4244-0431-2/06, IEEE, Pg 229.*

Wu KT, et. al., (2011). Engine oil condition monitoring using high temperature integrated ultrasonic transducers. *International Journal of Prognostics and Health Management*, 010, (ISSN 2153-2648).

Yates, S.W., Mosleh, M., (2006). A Bayesian approach to reliability demonstration for aerospace systems. *Proc. IEEE Explore, Reliability and Maintainability Symposium*, 611-617.

Ye Hua, Basaran, C., Hopkins, DC, (2006). Experimental damage mechanics of micro / power electronics solder joints under electric current stresses. International Journal of Damage Mechanics, Vol. 15, doi:10.1177/105678906054311.

Yin Chunyan, Lu Hua, MUSALLAM M., Bailey C. and Johnson C. Mark, (2008). Prognostic reliability analysis of power electronics modules. 2nd Electronics System integration Technology Conference, Greenwich, UK.

BIOGRAPHIES

Prabhakar V. Varde completed B.E.(Mech) from Government Engg College Rewa, in 1983, and joined BARC, Mumbai, in 1984 and worked with Reactor Operations Engineer till 1995. He obtained his Ph.D. from IIT, Bombay in 1996. He started R&D in the area of risk-based applications that includes development of PSA based expert system for nuclear plants. He worked as post-doctoral fellow at Korea Atomic Energy Research Institute, South Korea in 2002 and Visiting Faculty at University of Maryland in 2009. He was member of the PSA committee of Atomic Energy Regulatory, India till May 2012. He is also a member of the Board of Studies (Engg. Sciences), Homi Bhabha National Institute, India. His research interests are development of prognostic models for nuclear plants components in general and electronic components in particular. He worked as specialists / consultant to many international organizations which include, OECD/NEA (WGRISK) Paris, IAEA, Vienna, etc. He is founder of Society for Reliability and Safety based in India. He played central role in successfully organizing the International Conference on Reliability, Safety and Hazard in 2005 and 2010. He is one of the Chief Editors for International Journal of Life Cycle Reliability and Safety Engineering. He is on editorial board of many international journals in the area of reliability and safety. Based on his research and development work, he has published over 90 publications in journals and conferences which include 4 books / conference proceedings. Presently he is heading Safety Evaluation and Manpower Training Section at Research Reactor Services Division, BARC and Senior Professor, Homi Bhabha National Institute, Mumbai. He also worked as a Visiting Professor at CALCE, University of Maryland, USA in 2012.

Michael G. Pecht holds a MS in Electrical Engineering, and a MS and PhD in Engineering Mechanics from the University of Wisconsin at Madison. He is a Professional Engineer, an IEEE Fellow, an ASME Fellow, an SAE Fellow and an IMAPS Fellow. He has previously received the European Micro and Nano-Reliability Award for outstanding contributions to reliability research, 3M Research Award for electronics packaging, and the IMAPS William D. Ashman Memorial Achievement Award for his contributions in electronics reliability analysis. He served as chief editor of the IEEE Transactions on Reliability for eight years and on the advisory board of IEEE Spectrum. He is chief editor for Microelectronics Reliability and an associate editor for the IEEE Transactions on Components and Packaging Technology. Professor Michael G. Pecht is the founder and Director of CALCE (Center for Advanced Life Cycle Engineering) at the University of Maryland, which is funded by over 150 of the world's leading electronics companies at more than US$ 6M/year. He is also a Chair Professor in Mechanical Engineering and a Professor in Applied Mathematics at the University of Maryland. He has written more than twenty books on product reliability, development, use and supply chain management and over 400 technical articles. In 2008, he was awarded the highest reliability honor, the IEEE Reliability Society's Lifetime Achievement Award. In 2010, he received the IEEE Exceptional Technical Achievement Award for his reliability contributions in the area of prognostics and systems health management.

Battery Capacity Estimation of Low-Earth Orbit Satellite Application

Myungsoo Jun[1], Kandler Smith[2], Eric Wood[3], and Marshall .C. Smart[4]

[1,2,3] *National Renewable Energy Laboratory, Golden, CO, 80401, USA*
myungsoo.jun@nrel.gov, kandler.smith@nrel.gov, eric.wood@nrel.gov

[4] *Jet Propulsion Laboratory, California Institute of Technology, Pasadena, CA 91109, USA*
marshall.c.smart@jpl.nasa.gov

ABSTRACT

Simultaneous estimation of the battery capacity and state-of-charge is a difficult problem because they are dependent on each other and neither is directly measurable. This paper proposes a particle filtering approach for the estimation of the battery state-of-charge and a statistical method to estimate the battery capacity. Two different methods and time scales have been used for this estimation in order to reduce the dependency on each other. The algorithms are validated using experimental data from A123 graphite/LiFePO$_4$ lithium ion commercial-off-the-shelf cells, aged under partial depth-of-discharge cycling as encountered in low-earth-orbit satellite applications. The model-based method is extensible to battery applications with arbitrary duty-cycles.

1. INTRODUCTION

Health and lifetime uncertainty presents a major barrier to the deployment of lithium-ion (Li-ion) batteries in large-scale aerospace, electric vehicle, and electrical grid applications with stringent life requirements. In the satellite industry, for example, the high cost of launch and the inability to make repairs once in orbit dictate the use of mature battery technologies with conservative duty-cycles to reduce risk. If battery health could be precisely tracked on orbit, the duty-cycle might be tailored to best utilize the remaining life and maximize the value of the investment. Similar opportunities may exist for electric vehicles to maximize battery lifetime by intelligently selecting driving routes and charging strategies. Markets for used electric vehicles and batteries also require accurate battery health assessment to mature to their full potential.

The field of prognostics and health management offers general approaches for combining real-time measurements, models and estimation algorithms to track the health and predict the remaining lifetime of batteries (Sheppard, Wilmering, & Kaufman, 2009; Goebel, 2010). Relevant performance/health metrics for battery applications are available power and energy. These can be expressed in terms of battery internal resistance and amp-hour (Ah) capacity, respectively. Battery models are needed to relate capacity and resistance to the current, voltage, and temperature measurement signals available in real-time. For regular predictable duty-cycles such as in unmanned aerial vehicles (Goebel, Saha, Saxena, Celaya, & Christophersen, 2008), simple algebraic relationships between current and voltage may be sufficient. For uncertain duty-cycles such as for electric vehicles, a dynamic model of the current and voltage relationship is necessary. Dynamic models can be in the form of circuit analogs (Verbrugge & Koch, 2006; Plett, 2006), or reduced order physics-based models (Santhanagopalan, Zhang, Kumaresan, & White, 2008; Smith, Rahn, & Wang, 2007; Smith, 2010; J. L. Lee, Chemistruck, & Plett, 2012). Physics-based approaches remain their own active subject of research and thus the simpler circuit model is applied in this work.

State-of-charge (SOC) is usually formulated as a reference model state and can be estimated by using various state estimation methods such as extended Kalman filter (Plett, 2004; J. Lee, Nam, & Cho, 2007; Charkhgard & Farrokhi, 2010; Kim & Cho, 2011; Hu, Youn, & Chung, 2012), unscented Kalman filter (Plett, 2006; Sun, Hu, Zou, & Li, 2011) or cubature Kalman filter (Chen, 2012). Those SOC estimation methods work well in certain situations but would not perform properly in other situations. Extended Kalman filters are prone to linearization errors and both extended Kalman filters and unscented Kalman filters are limited to systems with Gaussian noise distribution. Similar to Kalman filters, particle filters belong to the class of Bayesian estimation methods, but can deal with nonlinear systems with non-Gaussian noise without linearization (Sanjeev Arulampalam, Maskell, Gor-

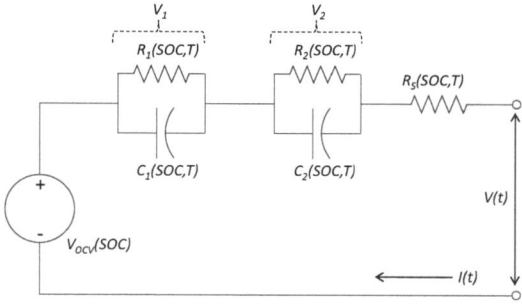

Figure 1. Second order circuit model of a battery

don, & Clapp, 2002). They have been successfully applied to many problems with nonlinear dynamics such as computer vision (Isard & Blake, 1998), speech recognition (Vermaak, Andrieu, Doucet, & Godsill, 2002), robotics (Schulz, Burgard, Fox, & Cremers, 2001), etc. Furthermore, very little work has been done in SOC estimation in conjunction with simultaneous estimation of time-varying battery capacity. This paper proposes a method to estimate both SOC and battery capacity by using a particle filtering approach.

Unlike in the laboratory, in an application environment it is infeasible to completely discharge the battery to obtain a full "ground-truth" measurement of battery total capacity. A key question explored in this paper is to what extent battery total amp-hour (Ah) capacity can be estimated based on only partial discharge data. In addition, estimation of battery capacity using partial discharge data is particularly challenging for Li-ion chemistries with a flat open-circuit voltage relationship versus SOC (Plett, 2011). Such is the case for the Li-ion graphite/iron-phosphate chemistry investigated in the present work.

2. CIRCUIT MODEL

For the reference model, a second-order circuit model is used in this work as shown in Figure 1. While the battery is an infinite-dimensional system, the two time constants of the second order circuit model provide reasonable approximation of voltage/current dynamics for the present application. The state-space equation of this circuit model is expressed as follows:

$$
\begin{bmatrix} \dot{SOC}(t) \\ \dot{V}_1(t) \\ \dot{V}_2(t) \end{bmatrix} = \begin{bmatrix} 0 & 0 & 0 \\ 0 & -\frac{1}{R_1 C_1} & 0 \\ 0 & 0 & -\frac{1}{R_2 C_2} \end{bmatrix} \begin{bmatrix} SOC(t) \\ V_1(t) \\ V_2(t) \end{bmatrix}
$$
$$
+ \begin{bmatrix} -\frac{1}{Q} & -\frac{1}{C_1} & -\frac{1}{C_2} \end{bmatrix}^T I(t) + n_1(t) \quad (1)
$$
$$
V_{out}(t) = V_{ocv}(SOC) - V_1(t) - V_2(t) -
$$
$$
R_s \cdot I(t) + v(t). \quad (2)
$$

where Q denotes the battery capacity. The values of the parameters R_1, R_2, R_s, C_1 and C_2 depend on SOC and time and Q depends on time. (Since the satellite battery considered in this work operates under nearly isothermal conditions, temperature dependency is neglected.)

Measurements of resistance versus SOC exhibit a bathtub shape, with small resistance at mid-SOCs increasing to larger values at low and high SOC extremes. This parametric dependence of R_1, R_2 and R_s on SOC is captured in Eq. (3)-(5)

$$
R_1(SOC(t)) = a_{r1} \left(1 + b_{r1} \cdot |SOC - c_{r1}|^{d_{r1}} \right) \quad (3)
$$
$$
R_2(SOC(t)) = a_{r2} \left(1 + b_{r2} \cdot |SOC - c_{r2}|^{d_{r2}} \right) \quad (4)
$$
$$
R_s(SOC(t)) = a_{rs} \left(1 + b_{rs} \cdot |SOC - c_{rs}|^{d_{rs}} \right). \quad (5)
$$

As the battery ages, the values of Q slowly decrease and the resistance values slowly increase over time. Since the battery may not be exercised over its entire SOC range in an actual application, only the three relative resistance parameters a_{r1}, a_{r2} and a_{rs} are estimated along with the battery capacity. The dynamics of these time-varying parameters can be formulated by:

$$
\begin{bmatrix} Q(k+1) \\ a_{r1}(k+1) \\ a_{r2}(k+1) \\ a_{rs}(k+1) \end{bmatrix} = M \begin{bmatrix} Q(k) \\ a_{r1}(k) \\ a_{r2}(k) \\ a_{rs}(k) \end{bmatrix} + n_2,
$$
$$
M = \begin{bmatrix} 1 - \varepsilon_1 & 0 & 0 & 0 \\ 0 & 1 + \varepsilon_2 & 0 & 0 \\ 0 & 0 & 1 + \varepsilon_2 & 0 \\ 0 & 0 & 0 & 1 + \varepsilon_2 \end{bmatrix} \quad (6)
$$

where ε_1 and ε_2 are small positive constants. We assume n_2 to be constant. We can reformulate a state-space equation by combining Eq.(1) and Eq. (6). Let $x(k) = [SOC(k) \ V_1(k) \ V_2(k) \ Q(k) \ a_{r1}(k) \ a_{r2}(k) \ a_{rs}(k)]^T$ be the augmented state and Δt be the sampling time. Then the discrete-time augmented state-space equation of the second-order circuit model of a battery is expressed as:

$$
x(k+1) = Ax(k) + BI(k) + n(k) \quad (7)
$$
$$
V_{out}(k) = V_{ocv}(SOC) - V_1(k) - V_2(k) -
$$
$$
R_s \cdot I(k) + v(k) \quad (8)
$$

where

$$
A = \text{diag}(1, e^{\lambda_1 \Delta t}, e^{\lambda_2 \Delta t}, M),
$$
$$
B = \begin{bmatrix} -\Delta t / Q \\ R_1(e^{\lambda_1 \Delta t} - 1) \\ R_2(e^{\lambda_2 \Delta t} - 1) \\ 0 \\ 0 \\ 0 \\ 0 \end{bmatrix},
$$
$$
n(k) = \text{diag}(n_1(k), n_2).
$$

3. PARTICLE FILTER

Particle filtering is a method used to approximate the probability density f_k of the state x_k conditioned on the observations $y_0, \cdots, y_k$[1]. Consider the following nonlinear system:

$$x_k = g(x_{k-1}, u_{k-1}) + n_k \tag{9}$$

$$y_k = h(x_k, u_k) + v_k. \tag{10}$$

where x_k is the state, y_k is the measurement, n_k is the process noise, and v_k is the measurement noise. Suppose that $f_{k-1} = p(x_{k-1} \mid y_0, \cdots, y_{k-1})$ is known. Then the *a priori* distribution of the state x_k can be derived via the Chapman-Kologorov equation[2]

$$p(x_k \mid y_0, \cdots, y_{k-1}) = \int p(x_k \mid x_{k-1}) f_{k-1} dx_{k-1} \tag{11}$$

where $p(x_k \mid x_{k-1})$ represents state transition over time and is determined by the process model (9) and the distribution of the process noise n_k. Note that f_{k-1} is a function of x_{k-1}. This step is called *prediction* or *time propagation*. When the observation y_k at time k is made, the *a priori* distribution is updated using Bayes' rule:

$$\begin{aligned} f_k &= p(x_k \mid y_0, \cdots, y_k) \\ &= \frac{p(y_k \mid x_k) \cdot p(x_k \mid y_0, \cdots, y_{k-1})}{\int p(y_k \mid x_k) \cdot p(x_k \mid y_0, \cdots, y_{k-1}) dx_k} \\ &= \frac{p(y_k \mid x_k) \cdot p(x_k \mid y_0, \cdots, y_{k-1})}{p(y_k \mid y_0, \cdots, y_{k-1})}. \end{aligned} \tag{12}$$

This step is called the *measurement update* as the measurement data $y_0, \cdots, y_k$ are used to obtain the *a posteriori* distribution $f_k = p(x_k \mid y_0, \cdots, y_k)$. The distribution $p(y_k \mid x_k)$ can be obtained from the measurement equations (10) and the distribution of the measurement noise v_k.

Particle filters approximate f_k by a set of weighted samples or particles x_k^i, $i = 1, \cdots, N$, where N is the number of particles. For more details about particle filters and sequential Monte Carlo methods, refer to (Sanjeev Arulampalam et al., 2002).

In this paper, sampling importance resampling is used for resampling of the particle filter to reduce degeneracy. The algorithm for the particle filter used in the simulations is given in the following:

1. Initialization: $k = 0$
   - Draw x_0^i, $i = 1, \cdots, N$ from the initial prior $f_0 = p(x_0)$.
2. For $k = 1, 2, \cdots$
   (a) Importance Sampling Step

- *State transition:* For $i = 1, \cdots, N$, draw x_k^i from $p(x_k^i \mid x_{k-1}^i)$, viz., from Eq. (9).
- *Measurement update and likelihood calculation:* For $i = 1, \cdots, N$, evaluate likelihood by calculating $w_k^i = p(y_k \mid x_k^i)$ after the measurement y_k is available.
- *Normalization:* Normalize the importance weights $\tilde{w}_k^i = w_k^i / \sum_j w_k^j$.

(b) Resampling

- Resample $\hat{x}_k^i$ using updated weights $\tilde{w}_k^i$.
- Set a new weight $w_k^i = 1/N$ for $i = 1, \cdots, N$.

4. CAPACITY ESTIMATION

The simultaneous estimation of the battery capacity and SOC is difficult because they are dependent on each other by the relation

$$SOC(t) = SOC(0) - \int_0^t \frac{I(\tau)}{Q} d\tau. \tag{13}$$

Therefore, if the changes in the battery capacity Q are not reflected properly, the calculation of SOC based on Eq. (13) is subject to errors even though the measurement of $I(t)$ is accurate. This paper proposes a novel method to estimate the battery capacity and SOC simultaneously using a particle filter and statistical approach.

The actual value of Q in real situations changes very slowly over time. This paper utilizes past statistical information for an estimate of Q at a longer interval than the sampling time. Let $m \gg 1$ be an integer and $T = m\Delta t$. The battery capacity is estimated at every T and the value of Q in Eq. (1) is set to the estimated battery capacity $\hat{Q}$ at every T other than Δt.

The estimate of x_k by the particle filter is the weighted sample mean of the particles, that is, $\hat{x}_k = \sum_{i=1}^N w_k^i x_k^i$ and the i-j-th element $q_k(i, j)$ of the weighted covariance matrix $\mathbf{Q}_k$ are

$$q_k(i, j) = \frac{\sum_{n=1}^N w_k^n}{\left(\sum_{n=1}^N w_k^n\right)^2 - \sum_{n=1}^N (w_k^n)^2} \times \\ \sum_{n=1}^N w_k^n \left(x_k^n(i) - \hat{x}_k(i)\right)\left(x_k^n(j) - \hat{x}_k(j)\right) \tag{14}$$

where $x_k^n(i)$ and $\hat{x}_k(i)$ are the i-th elements of the vectors x_k^n and $\hat{x}_k(i)$, respectively. The value of $q_k(4, 4)$ implies an estimation error for $x_k(4) = Q(k)$ and the degree of confidence can be represented by the reciprocal of $q_k(4, 4)$. Thus, the paper uses as estimate of Q

$$\hat{Q}(\ell T) = \sum_{k=(\ell-1)T}^{\ell T} \frac{W_k}{\sum_{j=(\ell-1)T}^{\ell T} W_j} \sum_{i=1}^N w_k^i x_k^i(4),$$
$$\ell = 1, 2, \cdots \tag{15}$$

---

[1] Let the subscript k denote discrete time k for simple notation, i.e., $x_k = x(k)$

[2] $p(x|y)$ means $p(X = x|Y = y)$ for simplicity of notation where X and Y are random variables and x and y are their realizations.

a_{rs}	b_{rs}	c_{rs}	d_{rs}	a_{r1}	b_{r1}	c_{r1}	d_{r1}	a_{r2}	b_{r2}	c_{r2}	d_{r2}
0.0105	112.8616	0.5221	5.5892	0.0440	0.0176	0.6307	3.5007	0.0157	0.0234	0.0029	7.3141

Table 1. Beginning of life parameter values in Eq.(3)-(5) for R_1, R_2 and R_s

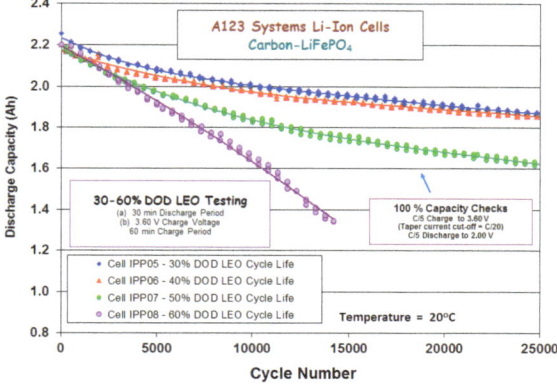

Figure 2. Capacity loss during partial DOD cycling of A123 LiFePO$_4$-based cells (Courtesy Jet Propulsion Laboratory)

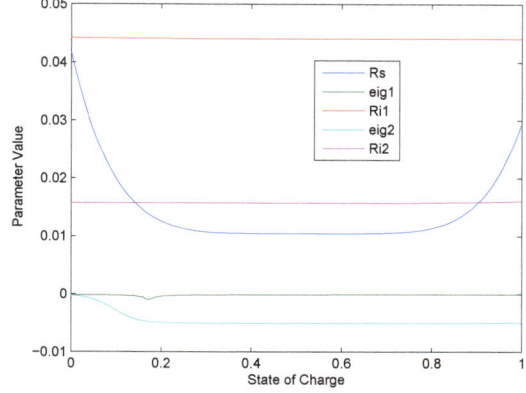

Figure 3. The values of resistances and eigenvalues identified at beginning of life by minimizing the square of voltage error between model and HPPC test data

where $W_k = 1/q_k(4,4)$ and the value of $Q(\ell T)$ is reset to a new value of Q in Eq. (1) for every ℓT, $\ell = 1, 2, \cdots$. This formulation can be interpreted that W_k is a weight and Eq (15) a weighted time average and re-initialization of state variables.

5. RESULTS

5.1. Low Earth Orbit Satellite Application

For the simulations, we used battery data generated at the Jet Propulsion Laboratory. They performed experiments to evaluate the cycle life performance of A123's 26650 LiFePO$_4$-based commercial off-the-shelf cells for potential low earth orbit satellite applications. This testing consists of implementing partial depth-of-discharge (DOD) cycling, with 30%, 40%, 50%, and 60% DOD selected. The testing was performed at the room temperature (23°C) and consisted of a 30-minute discharge period and a 60-minute charge period. The charge and discharge rates were scaled proportionately to the corresponding DOD (i.e., the 30% DOD test involved using a 0.4C charge rate and a 0.60C discharge rate). For operational capacity checks (OPCAPS), full charge and discharge of the battery were conducted every 250 cycles. The plots of battery capacity with respect to cycle number are shown in Figure 2. The degradation of battery capacity is clearly observed from the plot.

The analysis contained in this paper focuses upon the 50% DOD data from cycle 2723 to 2815. The battery capacity is reduced to about 2.05 Ah from the initial 2.2 Ah in the range of these cycles.

Least-square regression was used to provide an initial set

of parameters representing the battery at beginning of life. Specifically, the Nelder-Mead algorithm (MATLAB function `fminsearch`) minimized the square of the error between the model's output voltage and measured voltage from a hybrid pulse power characterization (HPPC) test run on a cell at beginning of life (Danzer & Hofer, 2008). The beginning-of-life values parameterizing R_1, R_2 and R_s as functions of SOC are shown in Table 1 and the plots of each resistance and eigenvalue are illustrated in Figure 3. Several nonlinearities arise in the model. Values of open-circuit voltage, $V_{ocv}(SOC)$ in Eq. (2), were taken at 10% increments in SOC following each one-hour rest period of the HPPC test and were implemented in the model as a look-up table. The nonlinearity in Eq. (1) lies in time-varying parameters, R_1, R_2, and R_s, which are also dependent on SOC.

The values of the parameters in the particle filter were tuned with simulations. We set the vales of ε_1 and ε_2 to be 0.00001. The value of the measurement noise v changes adaptively depending on SOC

$$v(SOC) = \begin{cases} v_0(1 + m_v(0.1 - SOC)), & 0.1 > SOC \\ v_0(1 + m_v(SOC - 0.9)), & 0.9 < SOC \\ v_0, & \text{otherwise} \end{cases}$$

where m_v is a scaling constant depending on measurement error when the value of SOC is very high or low. The process noise n_2 is set to be a constant and n_1 to be a function of SOC

$$n_1(SOC) = \begin{cases} n_{10}, & SOC \geq 0.1 \\ n_{10}(1 + m_w(0.1 - SOC)), & SOC < 0.1 \end{cases}$$

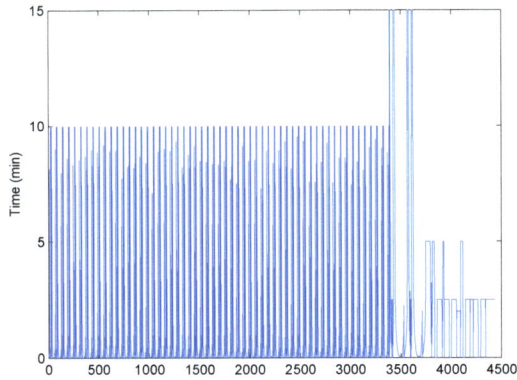

Figure 4. The time intervals between data samples from JPL experiment data

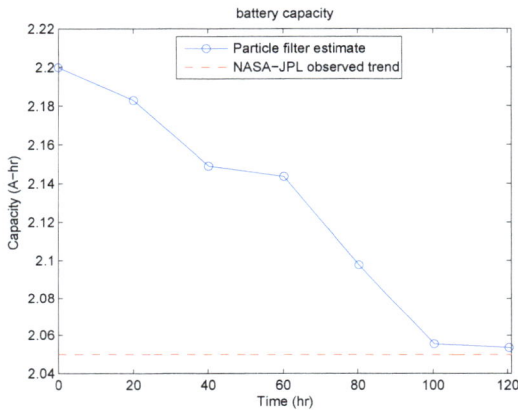

Figure 5. Battery capacity estimation with OPCAPS

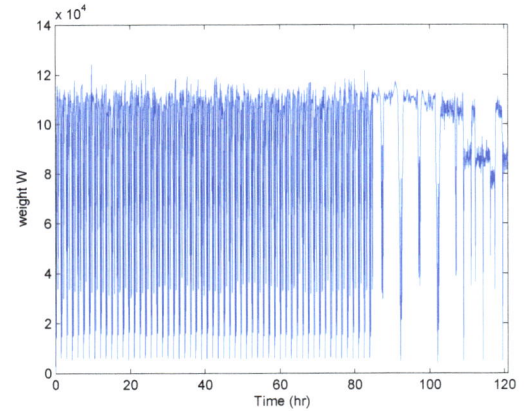

(a) Weight W_k

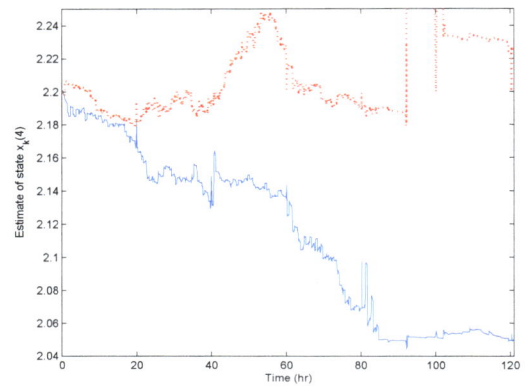

(b) Estimate of $x_k(4)$

Figure 6. Weight W_k and estimate of $x_k(4) = \sum_{i=0}^{N} w_k^i x_k^i(4)$

where m_w is a scaling constant.

The number of particles used in the simulations is 3,000. The sampling time of the filter is dependent on the interval of measurements. The plot of measurement intervals that are used for the simulations in Section 5.1.1 is shown in Figure 4. Measurements were mostly sampled at every 10 minute and the biggest sampling interval is 15 minute. The particle filter used in the simulations performs stratified resampling (Kitagawa, 1996) if

$$\widehat{N_{eff}} = \frac{1}{\sum_{i=1}^{N}(w_k^i)^2} < 0.5N$$

where N is the number of particles. Otherwise, the particle filter resamples using the normalized importance weight described in Section 3.

5.1.1. Simulation with Operational Capacity Checks (OP-CAPS)

First, we performed simulations with the data from cycle 2773 to 2815 that include OPCAPS. The estimate of the battery capacity was done every 20 hours, that is $T = 20$ hr in Eq. (15). Figure 5 shows the plot of the battery capacity estimate. The initial value of Q is set to 2.2 at time 0, which is the initial battery capacity before battery degradation. This initial value is kept until 20 hr. At $T = 20$ hr, the battery capacity is estimated to be about 2.1831 and the state variable Q in the particle filter is re-initialized to this value, and so on.

The plots of weight $W_k = 1/q_k(4,4)$ and the estimate of $x_k(4) = \sum_{i=0}^{N} w_k^i x_k^i(4)$ that are used for the battery capacity estimation using Eq. (15) are shown in Figure 6a. Figure 6b presents estimation of state $x_k(4)$ with (solid line) and without (dotted line) the proposed two-time scale method, respectively. It demonstrates that the proposed two-time

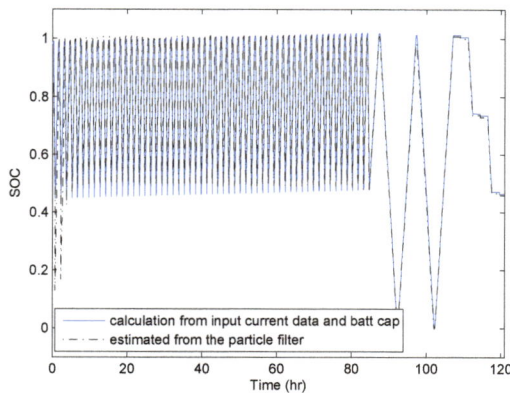

(a) SOC estimation

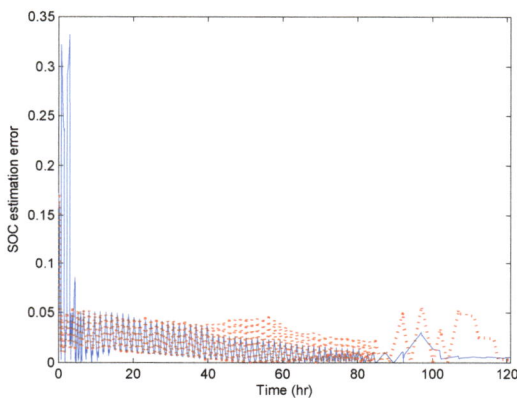

(b) Estimation error

Figure 7. Battery SOC estimation and error with OPCAPS

scale method performs better than particle filtering using augmented state which is usually used for the simultaneous estimation of state and parameters.

The SOC estimate and estimation error are illustrated in Figure 7. The blue solid line in Figure 7a shows the SOC value calculated from input current data and the true battery capacity (2.05 Ah) by using Eq. (13) and the black dash-dotted line illustrates estimated SOC from the particle filter. It can be observed in this plot that the peak value of SOC calculated from Eq. (13) increases over time. This is because measurement errors are accumulated through integration in Eq. (13) and the estimation using the particle filter is more robust to the measurement errors. The estimation error in Figure 7b is the difference between the estimate by the particle filter and the calculated value from input current data and the true battery capacity (2.05 Ah) using Eq. (13). The solid line in Figure 7b indicates SOC estimation error by using the proposed two-time scale method and the dotted line represents error by particle filtering without two-time scale. It shows

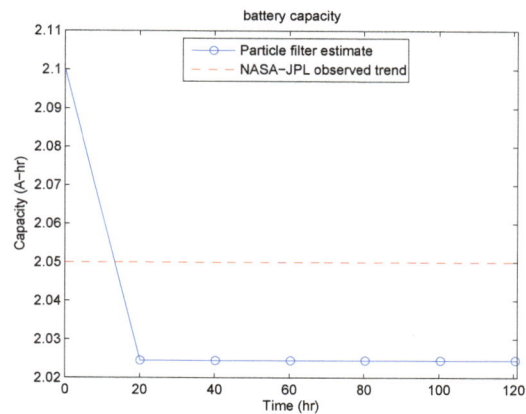

Figure 8. Battery capacity estimation without OPCAPS

that the error goes below 0.01 (1%) after about 100 hr by using the proposed method while the error does not decrease without two-time scale method.

5.1.2. Simulation without Operational Capacity Checks (OPCAPS)

The second simulation was performed with the data from cycle 2723 to 2806, which does not include OPCAPS and only has repeated charge and discharge with 50% DOD. The simulation results are shown in Figure 8, 10 and 9. In this case, the accumulated error in the SOC calculation by Eq. (13) is more noticeable. However, the SOC estimation using a particle filter oscillates between 0.5 and 1, which is the expected SOC range with 50% DOD.

The estimate of the battery capacity converges to about 2.025 Ah, which is a little less than 2.05 Ah, the actual capacity. The errors in SOC and the battery capacity estimation without OPCAPS are greater than those with OPCAPS and it took longer time to converge for SOC estimation. However, the error is about 1.2% and the estimate can be concluded to be accurate even without OPCAPS.

6. CONCLUSION

A method to simultaneously estimate both the capacity and SOC of a Li-ion battery has been proposed using a particle filtering method for SOC estimation and a statistical approach for the battery capacity. The battery capacity estimation has been performed in a different time scale from the SOC estimation and used accumulated past data from both measurement and the particle filter outputs. The estimated value of the battery capacity has been used to modify the parameter of the battery state-space model. Simulation results showed the robust performance of the algorithm in simultaneous estimation with or without operational capacity checks. The proposed method has been shown to perform better than the particle

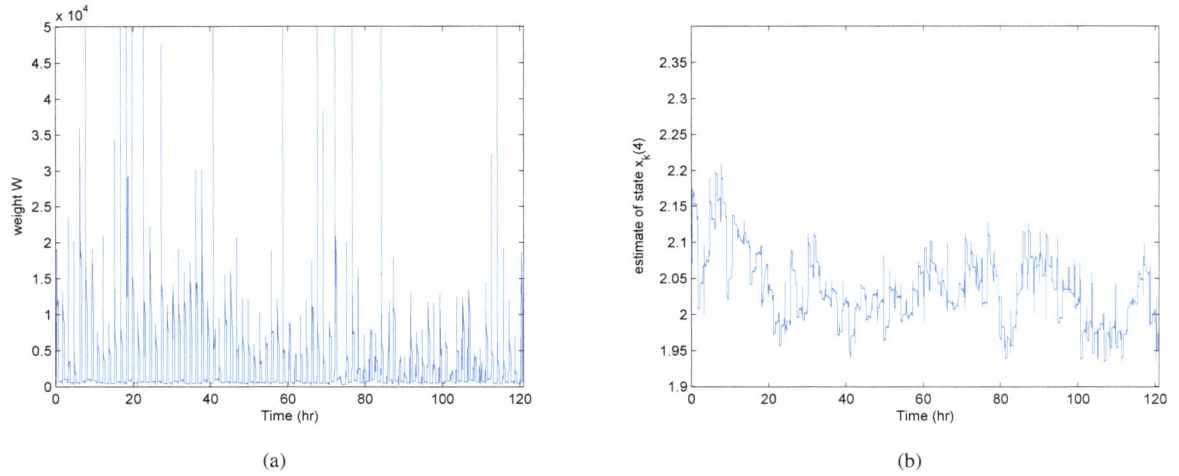

(a)

(b)

Figure 9. Weight W_k and estimate of $x_k(4) = \sum_{i=0}^{N} w_k^i x_k^i(4)$

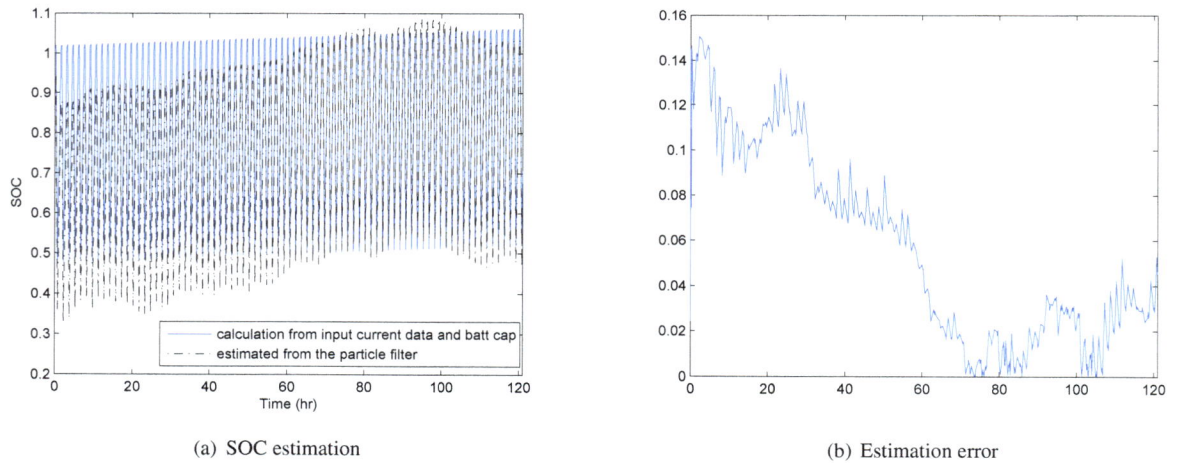

(a) SOC estimation

(b) Estimation error

Figure 10. Battery SOC estimation and error without OPCAPS

filter with one time scale and augmented state. Furthermore, accumulation of bias error over time has been shown to be corrected in SOC estimation with the proposed method.

Due to the high cost of launch, satellite batteries are expected to operate until the end of the satellite's life. Unlike the OP-CAP used in laboratory tests, in space the battery can never be fully discharged and hence the battery's total capacity must be indirectly estimated. Trending of battery total capacity over lifetime is important for satellite health management to ensure that no regular partial discharge cycle ever exceeds the present capability of the battery, causing loss of the satellite. The proposed method is adequate for the satellite applications since it estimates the battery capacity and SOC robustly even without OPCAPS and measurement errors are not accumulated in SOC estimation unlike Coulomb count method, which indicates that it is suitable for the applications with long operation time.

ACKNOWLEDGMENT

M. Jun, K. Smith and E. Wood gratefully acknowledge the National Renewable Energy Laboratory Laboratory-Directed Research and Development program for battery health algorithm development. The work performed by M. C. Smart was carried out at the Jet Propulsion Laboratory, California Institute of Technology, under contract with National Aeronautics and Space Administration (NASA).

REFERENCES

Charkhgard, M., & Farrokhi, M. (2010). State-of-Charge Estimation for Lithium-Ion Batteries Using Neural Networks and EKF. *IEEE Trans. Industrial Electronics*, *57*(12), 4178-4187.

Chen, H. (2012, july). Adaptive Cubature Kalman Filter for Nonlinear State and Parameter Estimation. In *Proc. of 15th International Conference on Information Fusion.* Singapore.

Danzer, M. A., & Hofer, E. P. (2008). Electrochemical parameter identification — An efficient method for fuel cell impedance characterization. *J. Power Sources*, *183*, 55-61.

Goebel, K. (2010, October). System Health Management: Predicting Failure after Fielding a Product. In *IEEE Accelerated Stress Testing and Reliability Workshop.* Denver, CO.

Goebel, K., Saha, B., Saxena, A., Celaya, J. R., & Christophersen, J. (2008, August). Prognostics in Battery Health Management. *IEEE Instrumentation & Measurement Magazine*, *11*(4), 33-40.

Hu, C., Youn, B. D., & Chung, J. (2012). A multiscale framework with extended Kalman filter for lithium-ion battery SOC and capacity estimation. *J. of Applied Energy*, *92*, 694-704.

Isard, M., & Blake, A. (1998). CONDENSATION — conditional density propagation for visual tracking. *International Journal of Computer Vision*, *29*(1), 5-28.

Kim, J., & Cho, B. H. (2011). State-of-Charge Estimation and State-of-Health Prediction of A Li-Ion Degraded Battery Based on An EKF Combined with A Per-Unit System. *IEEE Trans. Vehicular Technology*, *60*(9), 4249-4260.

Kitagawa, G. (1996). Monte-Carlo filter and smoother for non-Gaussian nonlinear state space model. *J. Comput. Graph. Statist.*, *1*, 1-25.

Lee, J., Nam, O., & Cho, B. H. (2007). Li-ion battery SOC estimation method based on the reduced order extended Kalman filtering. *J. of Power Sources*, *174*, 9-15.

Lee, J. L., Chemistruck, A., & Plett, G. L. (2012, December). One-dimensional physics-based reduced-order model of lithium-ion dynamics. *J. Power Sources*, *220*, 430448.

Plett, G. L. (2004). Extended Kalman filtering for battery management systems of LiPB-based HEV battery packs. Part 3: State and parameter estimation. *J. of Power Sources*, *134*(2), 277-292.

Plett, G. L. (2006). Sigma-point Kalman filtering for battery management systems of LiPB-based HEV battery packs. Part 2: Simultaneous state and parameter estimation. *J. of Power Sources*, *161*, 1369-1384.

Plett, G. L. (2011). Recursive approximate weighted total least squares estimation of battery cell total capacity. *J. of Power Sources*, *196*, 2319-2331.

Sanjeev Arulampalam, M., Maskell, S., Gordon, N., & Clapp, T. (2002). A tutorial on particle filters for online nonlinear/non-Gaussian Bayesian tracking. *IEEE Trans. on Signal Processing*, *50*(2), 174-188.

Santhanagopalan, S., Zhang, Q., Kumaresan, K., & White, R. E. (2008). Parameter estimation and life modeling of lithium-ion cells. *J. Electrochem. Soc.*, *155*(4), A345-A353.

Schulz, D., Burgard, W., Fox, D., & Cremers, A. (2001). Tracking multiple moving targets with a mobile robot using particle filters and statistical data association. In *Proc. of the IEEE International Conference on*

Robotics and Automation (p. 1665-1670). Seoul, Korea.

Sheppard, J. W., Wilmering, T. J., & Kaufman, M. A. (2009). IEEE Standards for Prognostics and Health Management. *IEEE Aerospace and Electronic Systems Magazine, 24*(9), 34-41.

Smith, K. (2010, April). Electrochemical Control of Lithium-Ion Batteries. *IEEE Control Systems Magazine.*

Smith, K., Rahn, C. D., & Wang, C. Y. (2007). Control-Oriented 1D Electrochemical Model of Lithium Ion Battery. *Energy Conversion and Management, 48*(9), 2565-2578.

Sun, F., Hu, X., Zou, Y., & Li, S. (2011). Adaptive unscented Kalman filtering for state of charge estimation of a lithium-ion battery for electric vehicles. *J. of Energy, 36*, 3531-3540.

Verbrugge, M. W., & Koch, B. J. (2006). Generalized recursive algorithm for adaptive multiparameter regression. *J. Electrochem. Soc., 153*(1), A187-A201.

Vermaak, J., Andrieu, C., Doucet, A., & Godsill, S. (2002, March). Particle methods for Bayesian modeling and enhancement of speech signals. *IEEE Trans. on Speech and Audio Processing, 10*, 173-185.

Sequential Monte Carlo methods for Discharge Time Prognosis in Lithium-Ion Batteries

Marcos E. Orchard[1], Matías A. Cerda[1], Benjamín E. Olivares[1], and Jorge F. Silva[1]

[1]*Department of Electrical Engineering, Universidad de Chile, Santiago, 8370451, Chile*

morchard@ing.uchile.cl
matias.cm.uc@gmail.com
benjamin.olivares.r@gmail.com
josilva@ing.uchile.cl

ABSTRACT

This paper presents the implementation of a particle-filtering-based prognostic framework that allows estimating the state-of-charge (SOC) and predicting the discharge time of energy storage devices (more specifically lithium-ion batteries). The proposed approach uses an empirical state-space model inspired in the battery phenomenology and particle-filtering to study the evolution of the SOC in time; adapting the value of unknown model parameters during the filtering stage and enabling fast convergence for the state estimates that define the initial condition for the prognosis stage. SOC prognosis is implemented using a particle-filtering-based framework that considers a statistical characterization of uncertainty for future discharge profiles.

1. INTRODUCTION

Energy storage devices (ESDs) play a key role in both industrial and military machinery. Consider, for example, electronic products commonly associated with daily living (laptop computers, communications equipment, GPS, domestic robots) and some other more sophisticated pieces of equipment and machinery such as pacemakers ground and aerial vehicles (manned, tele-commanded and unmanned), and satellites. ESDs may be used as primary sources of energy or as backup, allowing energizing different devices and systems under various operating profiles; thus improving their autonomy.

Particularly nowadays, Lithium-based compounds have presented significant advantages over other chemical combinations, such as Ni-MH, Ni-Cd and lead, when used for the manufacture of ESDs. In part this is explained by the fact that Li-Ion ESDs offer larger charge density by unit of mass (or volume), allowing to design very compact cells that can be easily integrated to small electronic devices; in addition, Li-Ion batteries offer extended life cycles and limited self-discharge rates (Saha and Goebel, 2009; Ranjbar *et al.*, 2012). Due to the exponential increase in the use of Li-Ion ESDs within the automotive industry, and the projected demand associated to this type of vehicles, the concept of "Battery Management Systems" – BMS, (Pattipati *et al.*, 2011) – started to become more a necessity than a luxury. These systems have as main objective (i) to provide and maximize usage time (autonomy) that is associated to a discharge cycle, (ii) to reduce battery charging times, (iii) to maximize the number of operating cycles for the ESD, and (iv) real-time operation, adjusting to sudden changes in charge/discharge conditions. To achieve this, BMS must consider at least information about the battery "state-of-charge" – SOC (Pattipati *et al.*, 2011), the "state-of-health" – SOH (Pattipati *et al.*, 2011), and the "remaining useful life" – RUL (Orchard and Vachtsevanos, 2009) of cells within the pack. On the one hand, knowledge about the SOC will help to quantify the autonomy of the system (based on some assumptions on the future discharge profile). On the other hand, knowledge about the SOH or the RUL of specific cells within the battery pack could help to decide when to replace or recycle it.

This article focuses on the problem of estimating and predicting the SOC of a Li-Ion ESD. Given that several definitions for the concept of SOC can be found in literature, it is relevant to mention that this research effort defines the SOC as the remnant energy (measured as a percentage of the current maximum cell capacity) in the battery. The SOC is highly affected by charge/discharge rate, temperature, hysteresis effects, usage time and self-discharge (due to the internal resistance of the cell), which transforms this particular prognosis problem into a very challenging one. Three specific aspects have to be carefully considered when intending to implement an efficient (and effective) SOC prognosis approach: (i) how to model the battery, (ii) how to estimate the SOC in a nonlinear, non-observable system, and (iii) how to predict the impact of future discharge profiles in the evolution of SOC in time. Several research efforts aim at providing a solution for the first two open questions, using empirical, physicochemical, or electric models in conjunction with estimation techniques based on fuzzy logic (Salkind et al., 1999), neural networks (Charkhgard and Farrokhi, 2010), or Bayesian approaches such as the extended Kalman filter – EKF (Saha and Goebel, 2009; Vinh Do et al., 2009).

This work has selected a combination of an empirical state-space model inspired in the battery phenomenology and Bayesian filtering techniques to study the evolution of the SOC in time; adapting and learning the value of unknown model parameters during the filtering stage, and enabling fast convergence for the state estimates that define the initial condition for the SOC prognosis stage. Due to model nonlinearities and the existence of non-Gaussian noise sources, the proposed approach considers the use of sequential Monte Carlo methods in prognosis, a sub-optimal filtering technique also known as particle-filtering. SOC prognosis is implemented using a particle-filtering-based framework (Orchard and Vachtsevanos, 2009) that considers uncertainty in the future discharge profile by including several possible future scenarios in the computation of the discharge time probability density function (PDF).

The structure of the article is as follows. Section 2 presents a theoretical framework for the problem of SOC estimation and prognosis, as well as failure prognosis using sequential Monte Carlo methods. Section 3 focuses on the modeling aspects that are required to incorporate the impact of different discharge profiles on the battery SOC. Section 4 shows the obtained results for SOC prognosis when trying to estimate the discharge time of a Li-Ion battery that energizes a ground robot. Finally, Section 5 presents the main conclusions of this research effort.

2. THEORETICAL FRAMEWORK

2.1. State-of-Charge Estimation in Lithium-Ion Batteries

The state-of-charge provides an indicator of the system autonomy that directly depends on the remaining battery energy and the mission profile; a critical piece of information for the design of path planning/control strategies in autonomous vehicles. It is for this reason that the implementation of SOC estimation and prognostic algorithms (Saha and Goebel, 2009; Ranjbar et al., 2012; Pattipati et al., 2011; Salkind et al., 1999; Charkhgard and Farrokhi, 2010; Vinh Do et al., 2009; Ran et al., 2010; Cadar et al., 2009; Qingsheng et al., 2010; Di et al., 2011; Tang et al., 2011) is considered the first step towards online characterization of both the End-of-Discharge (EoD) time

and RUL of Li-Ion batteries. One of the main difficulties in SOC estimation is that it cannot be measured directly, and thus its value must be inferred from the observation of other variables such as the battery current, voltage, temperature, state-of-health degradation and self-discharge phenomena (Pattipati et al., 2011; Cadar et al., 2009; Qingsheng et al., 2010; Di et al., 2011). Indeed, the utilization of more complex electrochemical models has been only suitable for off-line studies, mainly because these models (i) require a large number of variables to represent the battery internal structure, (ii) assume extremely accurate measurements, and (iii) have an elevated computational cost (Pattipati et al., 2011; Charkhgard and Farrokhi, 2010). Other options for SOC monitoring include the open-circuit voltage (OCV) method. This approach has the advantage of providing a direct relationship between battery SOC and voltage measurements – the higher the OCV, the higher the SOC (Tang et al., 2011). Unfortunately, the implementation of this test requires large resting periods for the battery, limiting its use for online applications (Pattipati et al., 2011; Charkhgard and Farrokhi, 2010; Di et al., 2011; Tang et al., 2011). Similarly, the "Electrochemical Impedance Spectroscopy" (EIS) (Pattipati et al., 2011; Ran et al., 2010) is a noninvasive method that intends to provide a complete characterization of the battery internal equivalent circuit. However, the implementation of an EIS test requires the acquisition of costly equipment (generally found only at laboratory test sites), which severely limits its widespread use in practice (Dalal et al., 2011). It is for this reason that current research efforts for SOC estimation and prognostic algorithms have focused on approaches that are mostly based on empirical models that incorporate only critical phenomenological aspects of the process; i.e., the relationship between currents, voltages and temperatures of Li-Ion cells. Among these methods, it is worth mentioning those that are based on fuzzy logic, neural networks, and Bayesian approaches.

On the one hand, fuzzy logic models have been used for the SOC estimation either through the identification of equivalent circuit for the battery from EIS data or directly from voltage and current measurements (Salkind et al., 1999). Given that EIS data have proved to be very noisy in practice (Saha et al., 2009; Dalal et al., 2011), only the latter case represents a reasonable method for online SOC estimation and uncertainty characterization. However, even in that case, the problem of SOC prediction (related to battery prognosis) is still unresolved and mainly treated as a curve regression problem (which is insufficient for purposes of risk characterization). Neural networks have also been used to build a nonlinear relationship between battery measurements and the evolution of SOC in time (Pattipati et al., 2011; Charkhgard and Farrokhi, 2010; Qingsheng et al., 2010). These methods, however, do not provide an adequate representation for uncertainty in nonlinear systems and thus neither can they be used for risk quantification purposes.

On the other hand, recent years have seen a growing interest in the use of machine learning techniques (e.g., Hamming Networks (Lee *et al.*, 2011)) and stochastic filtring techniques (unscented Kalman Filter (Santhanagopalan and White, 2010), extended Kalman Filter (Hu *et al.*, 2012), and unscented particle filter (He *et al.*, 2013)) to estimate the SOC and/or parameter degradation of a Li-Ion battery cell under a randomly varying loading condition. Suboptimal Bayesian methods have proven particularly effective in the task of simultaneously incorporate information from noisy measurements and characterize the sources of uncertainty (Charkhgard and Farrokhi, 2010; Vinh Do *et al.*, 2009; Di *et al.*, 2011; Saha *et al.*, 2009; Dalal *et al.*, 2011; Orchard *et al.*, 2010). In fact, experience has demonstrated that Bayesian state estimators are especially well suited for real-time estimation problems associated to dynamic state models (Saha *et al.*, 2009; Dalal *et al.*, 2011; Orchard *et al.*, 2010). In addition, these methods also provide a concrete characterization of uncertainty sources both in the filtering and the prediction stage, a piece of information that is required for the generation of a risk measure associated to SOC prognosis. Bayesian estimators require a state-space model for the dynamic system, and prognostic modules based on a state-space formulation for the dynamic system are very sensitive to the initial condition of the state vector. For this reason, the implementation of accurate online SOC estimators is absolutely relevant for the development of real-time predictors capable of quantifying the feasibility (as well as the cost) of a particular vehicle trajectory. Depending on the validity of linear or Gaussian assumptions, either an extended Kalman filter or a particle-filtering (Arulampalam *et al.*, 2002; Andrieu *et al.*, 2001; Doucet *et al.*, 2001) approach may be needed.

To evaluate which is the best option for this particular problem, it is first necessary to define a state model that represents adequately the dynamics associated to the evolution of the battery SOC, for a given usage profile. This research has considered for this purpose the problem of battery end-of-charge prognosis in a four-wheel ground robot. In this scenario, different discharge profiles can be verified due to terrain conditions (hills, surface changes) and other factors, while the robot autonomously has to perform a pre-determined mission. Given that the robot is currently configured only for use on 2D, uniform terrain, it is necessary to simulate the environment through a variable load has been attached to the battery. This variable load is made up of three resistors (6.23Ω, 12.5Ω, and 25Ω), each wired in parallel to the battery to increase current draw. Each resistor can be activated via a relay controlled by the onboard computer's data acquisition card. It provides 8 different loading scenarios progressing linearly in magnitude. The onboard computer has a map of simulated terrain and when the robot crosses into an area of higher simulated difficulty on the map, the onboard computer activates a larger loading scenario using the variable load.

This allows for many simulated terrains while keeping the robot in a safe, uniformly flat environment. Online data consists of voltage and current measurements (with the corresponding timestamp), for a lithium iron phosphate ($LiFePO_4$) battery (12.8[V], 2.4[Ah], 14[A] maximum discharge current). This experimental setup implies that for full speed (700 [mm/s]), the current drained from the battery ranges between 1.6025[A] and 5.4738[A] (depending on the value of the equivalent resistor that is connected in parallel to the battery); at 10% speed (70[mm/s]) the current drained from the battery ranges between 0.6006[A] and 4.3971[A].

2.2. Particle-Filtering-based Prognosis Framework for Faulty Dynamic Nonlinear Systems

Consider a sequence of probability distributions $\{\pi_k(x_{0:k})\}_{k\geq1}$, where it is assumed that $\pi_k(x_{0:k})$ can be evaluated pointwise up to a normalizing constant. Sequential Monte Carlo (SMC) methods, also referred to as particle filters (PF), are a class of algorithms designed to approximately obtain samples from $\{\pi_k\}$ sequentially; i.e., to generate a collection of $N>>1$ weighted random samples (particles) $\{w_k^{(i)}, x_{0:k}^{(i)}\}_{i=1\cdots N}$, $w_k^{(i)} \geq 0, \forall k \geq 1$, satisfying (Andrieu *et al.*, 2001; Doucet *et al.*, 2001):

$$\sum_{i=1}^{N} w_k^{(i)} \varphi_k(x_{0:k}^{(i)}) \xrightarrow[N\to\infty]{} \int \varphi_k(x_{0:k}) \pi_k(x_{0:k}) dx_{0:k}, \qquad (1)$$

in probability and where φ_k is any π_k-integrable function.

In the particular case of the Bayesian Filtering problem, the *target distribution* $\pi_k(x_{0:k}) = p(x_{0:k} \mid y_{1:k})$ is the posterior PDF of $X_{0:k}$, given a realization of noisy observations $Y_{1:k} = y_{1:k}$.

Let a set of N paths $\{x_{0:k-1}^{(i)}\}_{i=1\cdots N}$ be available at time $k-1$. Furthermore, let these paths distribute according to $q_{k-1}(x_{0:k-1})$, also referred to as the importance density function at time $k-1$. Then, the objective is to efficiently obtain a set of N new paths $\{\tilde{x}_{0:k}^{(i)}\}_{i=1\cdots N}$ distributed according to $\pi_k(\tilde{x}_{0:k})$ (Andrieu *et al.*, 2001).

For this purpose, the current paths $x_{0:k-1}^{(i)}$ are extended by using the kernel $q_k(\tilde{x}_{0:k} \mid x_{0:k-1}) = \delta(\tilde{x}_{0:k-1} - x_{0:k-1}) \cdot q_k(\tilde{x}_k \mid x_{0:k-1})$; i.e., $\tilde{x}_{0:k} = (x_{0:k-1}, \tilde{x}_k)$. The importance sampling procedure generates consistent estimates for the expectations of any function, using the empirical distribution (Doucet *et al.*, 2001):

$$\tilde{\pi}_k^N(x_{0:k}) = \sum_{i=1}^{N} w_{0:k}^{(i)} \delta(x_{0:k} - \tilde{x}_{0:k}^{(i)}), \qquad (2)$$

where $w_{0:k}^{(i)} \propto w_{0:k}(\tilde{x}_{0:k}^{(i)})$ and $\sum_{i=1}^{N} w_{0:k}^{(i)} = 1$.

The most basic SMC implementation –the sequential importance sampling (SIS) particle filter– computes the value of the particle weights $w_{0:k}^{(i)}$, by setting the importance density function equal to the *a priori* state transition PDF $p(\tilde{x}_k \mid x_{k-1})$; i.e., $q_k(\tilde{x}_{0:k} \mid x_{0:k-1}) = p(\tilde{x}_k \mid x_{k-1})$. In that manner, the weights for the newly generated particles are evaluated from the likelihood of new observations. The efficiency of the procedure improves as the variance of the importance weights is minimized. The choice of the importance density function is critical for the performance of the particle filter scheme and hence, it should be considered in the filter design.

Prognosis (Engel *et al.*, 2000), and thus the generation of long-term prediction, is a problem that goes beyond the scope of filtering algorithms since it involves future time horizons. Hence, if PF-based algorithms (Orchard *et al.*, 2009; Edwards *et al.*, 2010; Chen *et al.*, 2011) are to be used, it is necessary to propose a procedure with the capability to project the current particle population in time in the absence of new observations.

Any adaptive prognosis scheme requires the existence of at least one feature providing a measure of the severity of the fault condition under analysis – fault dimension (Zhang *et al.*, 2011). If many features are available, they can always be combined to generate a single signal. In this sense, it is always possible to describe the evolution in time of the fault dimension through the nonlinear state equation.

By using the aforementioned state equation to represent the evolution of the fault dimension in time, it is possible to generate *p*-ahead long term predictions, using kernel functions to reconstruct the estimate of the state PDF in future time instants, as it is shown in Eq. (3):

$$\tilde{p}(x_{k+p} \mid \tilde{x}_{1:k+p-1}) \approx \sum_{i=1}^{N} w_{k+p-1}^{(i)} K\left(x_{k+p} - E\left[x_{k+p}^{(i)} \mid \tilde{x}_{k+p-1}^{(i)}\right]\right), \quad (3)$$

where $K(\cdot)$ is a kernel density function, which may correspond to the process noise PDF, a Gaussian kernel or a rescaled version of the Epanechnikov kernel (Orchard and Vachtsevanos, 2009; Orchard *et al.*, 2010; Orchard, Tobar and Vachtsevanos, 2009). The resulting predicted state PDF contains critical information about the evolution of the fault dimension over time. One way to represent that information is through the computation of statistics (expectations, 95% confidence intervals), either the End-of-Discharge (EOD) (Saha *et al.*, 2009) or the Remaining Useful Life (RUL) of the faulty system.

The EOD PDF depends on both long-term predictions and empirical knowledge about critical conditions for the system (Saxena *et al.*, 2010; Tang, Orchard *et al.*, 2011). This empirical knowledge is usually incorporated in the form of thresholds for main fault indicators. Therefore, the probability of failure at any future time instant $k = eod$

(namely the EOD PDF) is given by (Orchard and Vachtsevanos, 2009):

$$\Pr\{EOD = eod\} = \sum_{i=1}^{N} \Pr\left(Failure \mid X = \hat{x}_{eod}^{(i)}\right) \cdot w_{eod}^{(i)}. \quad (4)$$

The conditional probability of failure in Eq. (4) may be defined via the determination of hazard zones (Orchard and Vachtsevanos, 2009), either using historical data or knowledge from process operators. The simplest case is where the concept of "failure" implies the moment when the fault feature crosses a given threshold. In that case the probability of failure, conditional to the state, is equal to one if the state is exactly on the manifold that defines the threshold value.

3. STATE-SPACE MODEL FOR STATE-OF-CHARGE ESTIMATION IN ESDS

This article focuses on the development and implementation of a module for online discharge time prognosis in Li-Ion ESDs, using an adequate characterization of the future usage profile. For this, firstly it is necessary to model the effect that any arbitrary discharge current profile may have on the battery SOC. This model should also have a reduced number of parameters, allowing estimating and prognosticating in real-time in and accurate and precise manner. Because of these facts, physicochemical models were discarded as a feasible choice since their complexity (and the need of extremely precise off-line measurements) implied a high computational cost. Other options, such as electric equivalent circuits, with parameters that could be estimated from EIS measurements, were also discarded since (as it has been already mentioned) the implementation of such tests requires the acquisition of costly equipment, severely limiting its widespread use in practice (Dalal *et al.*, 2011). Therefore, this research effort has chosen a grey-box (Gonzalez *et al.*, 2003) empirical (discrete-time) model that is inspired in the battery phenomenology. The selected model only depends on voltage and discharge current measurements, something that enables its use for online prognostics modules. The proposed structure requires defining necessarily one state as the battery SOC. In addition, it is important to consider adaptation of the model to a particular battery under supervision, through the definition of a state associated to an unknown model parameter. The model used is shown below:

State transition model:

$$x_1(k+1) = x_1(k) + \omega_1(k), \quad (5)$$

$$x_2(k+1) = x_2(k) + v(k) \cdot i(k) \cdot \Delta t \cdot 10^{-5} + \omega_2(k), \quad (6)$$

Measurement equation:

$$v(k) = \left[v_0 - x_1(k) \cdot i(k) - e^{C \cdot x_2(k)} + \eta(k)\right], \quad (7)$$

$$v_0 = 13[V]; \quad C = 5.5687; \quad x_1(0) = 0.0897; \quad x_2(0) = -E_0$$

where the battery current level $i(k)$ [A] and the sampling period Δt [sec] are input variables, and the battery voltage $v(k)$ [V] is the system output. The states are defined as $x_1(k)$ (unknown model parameter) and $x_2(k)$ (additive inverse of SOC, remnant battery energy measured in [VA sec 10^{-5}]), E_0 is the ESD initial SOC (that could be inferred from data acquired during the charging process).

Process noises ω_1 and ω_2 represent uncertainty on the a priori state estimates, and C is a constant that characterizes the battery voltage drop in terms of the remaining SOC. It is important to note that process noise (at least noise ω_2) is correlated with η, the measurement noise, since uncertainty on battery SOC depend on the uncertainty of voltage measurements. This fact will be considered when designing the prognostic module. The state x_1 in Eq. (7) represents the instantaneous value for the battery internal resistance. It is well known that this value depends on other environmental factors (e.g., temperature). As in this case the experimental setup did not include temperature probes, then the filtering stage must infer the effect of the external temperature (and other unmeasured perturbations) into the state estimate x_1, based solely on voltage and current measurements.

Validation data (see Figure 1) for the implemented algorithms was obtained from a mobile platform developed at Impact Technologies, LCC, which basically consisted of a four-wheel ground robot used to generate different battery discharge profiles due to terrain conditions (hills, surface changes). Given that the robot is currently configured only for use on 2D, uniform terrain, it is necessary to simulate the environment through a variable load has been attached to the battery. This variable load is made up of three resistors (6.23[Ω], 12.5[Ω], and 25[Ω]), each wired in parallel to the battery to increase current draw. Each resistor can be activated via a relay controlled by the onboard computer's data acquisition card. It provides 8 different loading scenarios progressing linearly in magnitude. The onboard computer has a map of simulated terrain and when the robot crosses into an area of higher simulated difficulty on the map, the onboard computer activates a larger loading scenario using the variable load. This allows for many simulated terrains while keeping the robot in a safe, uniformly flat environment.

Online data consists of voltage and current measurements (with the corresponding timestamp), for a lithium iron phosphate (LiFePO$_4$) battery (12.8[V], 2.4[Ah], 14[A] maximum discharge current). As the battery voltage drops, maximum values of the drain current may increase. Figure 1 show measured data for the battery voltage [V] and current [A] in an experiment where the energy accumulator was used until it discharged almost completely.

The nonlinear model proposed in Eq. (5)-(7) simultaneously allows a statistical characterization of future battery discharge profiles and the implementation of Bayesian prognostic approaches, such as those based on particle-

filtering techniques (Orchard and Vachtsevanos, 2009; Saha et al., 2009; Orchard et al., 2010; Orchard, Tobar, and Vachtsevanos, 2009; Edwards et al., 2010; Chen et al., 2011; Zhang et al., 2011). Statistical characterization of discharge current profiles may incorporate information from past measurements, mainly from the ESD current $i(k)$, to understand the manner in which the ESD has been used lately. Aggressive usage profiles will translate in reduced discharge times and, conversely, low energy consumptions will lead to extended period of autonomy. Although diverse factors may affect future utilization of the ESDs, typically the system that uses those devices as primary sources of energy will operate trying to achieve similar performance indicators before and after the prognosis time; i.e., the time when long-term predictions are computed. For this reason it is critical to characterize this usage profile during the filtering stage, where the battery current $i(k)$ is an input to the estimator, and thus a known signal.

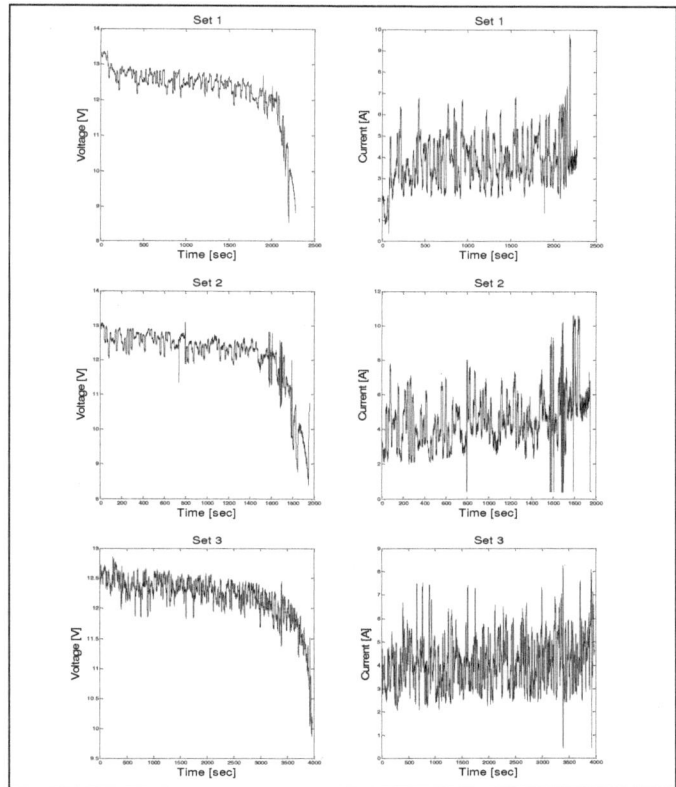

Figure 1: Measured voltage and discharge current data for a ground robot using a Li-Ion ESD. Curves on the left show voltage [V] as the battery discharges, while the curves on the right show the battery current [A].

The proposed approach for statistical characterization of battery discharge current profiles is based on a combination of algorithms that define an operating range through the computation of maximum/minimum discharge currents and empiric distributions of the acquired measurement data. The procedure used to compute extreme values is as follows:

1. Firstly, data from the current usage profile (measured battery current) is segmented in identical time intervals, as shown in Figure 2.

2. On each interval, the maximum and minimum values for the battery current, $i_{low}^{(m)}$ and $i_{high}^{(m)}$ respectively, are computed (m is an index for the mth interval). These values define a range for the battery current that could be used to characterize the future operation profile of the battery, in a scenario where the prognosticator solely considers data from the mth time interval to this purpose. A low-pass filter was used to discard outliers and/or anomalous currents peaks.

3. An exponentially weighted moving average – EWMA (Hunter, 1986) is used to reduce the impact of an arbitrary definition for time intervals, and to incorporate prior information about battery usage. EWMA computes the extreme values that will be used to characterize the range for future battery discharge currents:

$$\bar{i}_{low}^{(m)} = (1-\alpha) \cdot i_{low}^{(m)} + \alpha \cdot \bar{i}_{low}^{(m-1)} \quad \forall m, \qquad (8)$$

$$\bar{i}_{high}^{(m)} = (1-\alpha) \cdot i_{high}^{(m)} + \alpha \cdot \bar{i}_{high}^{(m-1)} \quad \forall m, \qquad (9)$$

where $\bar{i}_{low}^{(m)}$ and $\bar{i}_{high}^{(m)}$ are, respectively, the minimum and the maximum values that would be assumed for the future battery discharge profile, if the prediction were to be computed at the end of the mth interval (see Figure 2). The parameter α corresponds to the forgetting factor of the EWMA algorithm ($\alpha = 0.65$ in this case).

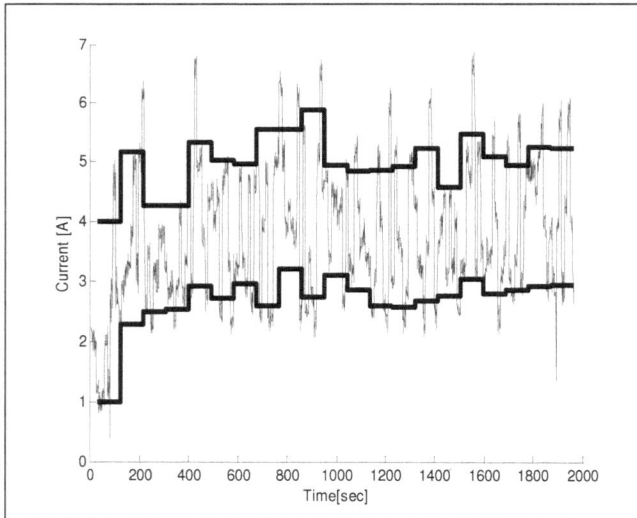

Figure 2: Characterization of maximum and minimum discharge current levels for future operation of the ESD based on exponentially weighted moving average algorithms.

The obtained range enables the generation of empirical distributions for the values that characterize the uncertainty in the future battery discharge profile; considering information about how the device has been used so far. More details on the construction of this empirical distribution will be described in the following section.

4. PARTICLE-FILTERING-BASED DISCHARGE TIME PROGNOSIS FOR LITHIUM-ION ESDs

The problem of battery EOD time has been deeply discussed by several authors in the recent years (see Section 2.1). Most of them, though, intend to learn the trend of the discharge curve assuming that the only sources of uncertainty are associated to unknown model parameters or the estimates of the state vector, while the future operating profile is assumed as a deterministic function of time (constant battery current, most of the times). This article proposes a framework that intends to complement the characterization of the uncertainty associated to future discharge profiles to improve the accuracy of prognostic results, combining a classic implementation of a particle-filtering-based prognostic framework (Orchard and Vachtsevanos, 2009; Orchard et al., 2010) with a statistical characterization of the previous battery usage.

Indeed, several results (see Section 2.1) indicate that Eq. (3)-(7) can be used to show that the a priori state PDF for future time instants, and thus the EOD PDF, directly depends on the a priori probability distribution of the battery discharge profile for future time instants. Most of the times, long-term predictions assume that the latter distribution is a Dirac's delta function (a deterministic function of time for future discharge profiles). Although this simplification helps to speed up the prognostic procedure and to generate the most likely EOD estimate, it does not allow considering future changes in operating conditions or unexpected events that could affect the autonomy of the system under analysis.

Monte Carlo simulation can be used to generate EOD estimates for arbitrary a priori distributions of future operating conditions, however it is not always possible to obtain these results in real-time. In this sense, PF-based prognostic routines not only provide a theoretical framework where these concepts can be incorporated in real-time (Edwards et al., 2010), but also allow the use of uncertainty measures to characterize the sensitivity of the system with respect to changes in future load distributions (see Figure 3). Furthermore, if a formal definition of mass probability is assigned to each possible operating condition, an EOD PDF estimate can be obtained as a weighted sum of kernels, where each kernel represents the PDF estimate of a known discharge current profile, characterized as a function of time. Indeed, if the a priori distribution of future operating conditions is given by:

$$\Pr\{U=u\}=\sum_{j=1}^{N_u}\pi_j\delta\left(u-u_j\right), \qquad (10)$$

where $\{u_j\}_{j=1}^{N_u}$ is a set of deterministic time functions, then the probability of failure at a future time t can be computed using Eq. (11) (Edwards $et\ al.$, 2010).

$$\Pr\{EOD=eod\}=\sum_{j=1}^{N_u}\pi_j\sum_{i=1}^{N}\Pr\left(Failure\mid X=\hat{x}_{eod}^{(i)},u=u_j\right)\cdot w_{eod}^{(i)}\cdot \qquad (11)$$

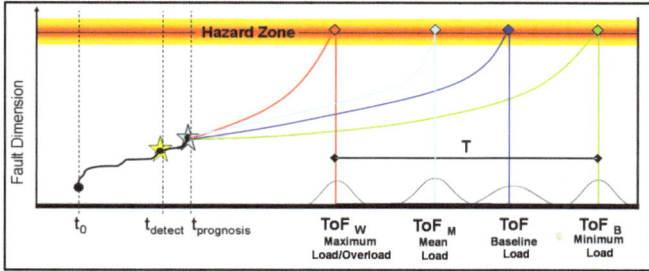

Figure 3: Illustration for the effect that different load/stress profiles may have on the growth of the fault dimension in a nonlinear dynamic system. More aggressive profiles typically result in a shorter RUL for the system.

This particular research effort considered a statistical characterization of future battery operation ($i(k)$, $k = t_{pred}+1, ..., EOD$) that assumed constant battery current profiles for each of the deterministic functions of time $\{u_j\}_{j=1}^{N_u}$. The probabilities π_j associated to each path u_j were computed through empirical distribution of past measurements for the battery current, and considering that the support (domain) of the empirical distribution is given by the interval $[\bar{i}_{low}^{(m)}, \bar{i}_{high}^{(m)}]$ that is calculated using Eq. (8)-(9).

Tables 1 through 3 show the obtained results when considering $N_u = 1$, 2, and 3 respectively, computed at a particular time instant where the long-term prediction is to be calculated for model (5)-(7). For each one of these cases, the value of the future battery discharge current u_j is computed as the center of an interval defined by the expression $[\bar{i}_{low}^{(m)}+j(\bar{i}_{high}^{(m)}-\bar{i}_{low}^{(m)})/N_u, \bar{i}_{low}^{(m)}+(j+1)(\bar{i}_{high}^{(m)}-\bar{i}_{low}^{(m)})/N_u]$, $j = 0,...,N_u$-1 In addition, once the empirical distribution is built, it is possible to compute the expectation of the future battery discharge current as a weighted average.

In this case, as the N_u increases, it is important to note that there is only a limited impact on the probability that is assigned to the intervals that represent maximum, minimum and the median discharge currents. This fact is critical when deciding the number of paths to be considered in long-term predictions (the battery), since each prediction path involves the computation of a conditional PDF in real-time (Edwards $et\ al.$, 2010). If limited computational resources are available, it may be wise to select the minimum possible number of paths that can help to represent the uncertainty of

the battery operating conditions. A throughout analysis of the data presented in Tables 1 through 3 indicates that three possible discharge profiles are sufficient to characterize the tails of the EOD PDF, since the EOD increases monotonically as the battery discharge current decreases.

j ($N_u = 3$)	# Samples	π_j	Current [A]
1st	228	0.4551	3.2965
2nd	132	0.2635	4.3177
3rd	141	0.2814	5.3388
Weighted Average	501	1	4.1403

Table 1: Empirical distributions that define the weights π_j for three future battery discharge profiles in a particle-filtering-based prognosis approach.

j ($N_u = 5$)	# Samples	π_j	Current [A]
1	184	0.3673	3.0923
2	82	0.1637	3.7050
3	58	0.1158	4.3177
4	115	0.2295	4.9303
5	62	0.1238	5.5430
Weighted Average	501	1	4.0596

Table 2: Empirical distributions that define the weights π_j for five future battery discharge profiles in a particle-filtering-based prognosis approach.

j ($N_u = 7$)	# Samples	π_j	Current [A]
1	171	0.3413	3.0048
2	42	0.0838	3.4424
3	71	0.1417	3.8800
4	31	0.0619	4.3177
5	78	0.1557	4.7553
6	57	0.1138	5.1929
7	51	0.1018	5.6305
Weighted Average	501	1	4.0355

Table 3: Empirical distributions that define the weights π_j for seven future battery discharge profiles in a particle-filtering-based prognosis approach.

Considering the aforementioned information, it was determined that the proposed approach for uncertainty representation would only include three possible paths within the implementation of the particle-filtering-based EOD prognosis framework. Lithium-Ion battery data shown in Section III was used to validate this method, comparing its performance with respect to a classic PF-based prognostic implementation (Saha and Goebel, 2009; Orchard and Vachtsevanos, 2009) that assumed only constant discharge current (computed as the average of past battery current measurements) for future operation.

Figure 4 and Figure 5 show the results obtained when using the classic PF-based prognostic approach at 920[sec] of operation (40 particles and using 25 realizations of the long-term prediction before computing the EOD PDF). On the one hand, Figure 4 illustrates the results of the filtering and prognostic stages, showing that indeed an implementation based on model (5)-(7) can be used to quantify the effect that random changes in the battery discharge current have on the voltage of the device. Moreover, the predicted battery voltage includes information about the effect of the future evolution of the battery SOC in time, exhibiting an exponential drop as the SOC decreases.

Similar conclusions can be obtained from Figure 5, where the prediction stage is emphasized. The resulting EOD PDF allows building a 95% confidence interval that provides information about when the battery would discharge if the future operating condition of the device is kept invariant.

Although this interval has a relative high precision (which is expected since there is no uncertainty associated to the future operating profile), it is important to mention that the bias associated to the assumed operation profile translated into overestimating the EOD time. Indeed, the computed expectation of the EOD time is 2224.91 [sec], while the ground truth discharge time occurred at 2123.7 [sec] of operation (101.21 [sec] before the expected value). A quick analysis of this information indicates that the proposed model for SOC estimation allows to efficiently incorporate measurement data and to generate reliable predictions paths; however, it must be said the high precision of the resulting PDF is only due to the fact that the uncertainty associated to the state estimates is bounded (the filter "learns" from data, thus the more extended the filtering stage is, the better the estimates) and because the uncertainty associated to the future battery use is neglected.

That is not necessarily true if the actual usage pattern of the battery is studied in detail. Figure 6 shows the obtained results when incorporating information from a statistical characterization, using the method proposed in Section 3 and Table 1. The fact of incorporating more uncertainty in the long-term predictions clearly has an impact on the precision of the resulting 95% confidence interval; however, as a trade-off, the accuracy of the EOD expectation (2137.5 [sec]) is highly increased (see Figure 7) to a point where the algorithm has an error of only 13.8 [sec] in a 1203.7 [sec] prediction window. Although at a first glance many researchers could feel tempted to indicate that the precision of the proposed approach is disappointing, one must remember that a good prognostic algorithm should correctly characterize all uncertainty sources. As it has been already mentioned, by neglecting the uncertainty associated to the usage profile it is obvious that the resulting EOD PDF will be more "precise". The real question is if the computed precision represents the manner in which the user is operating the actual system.

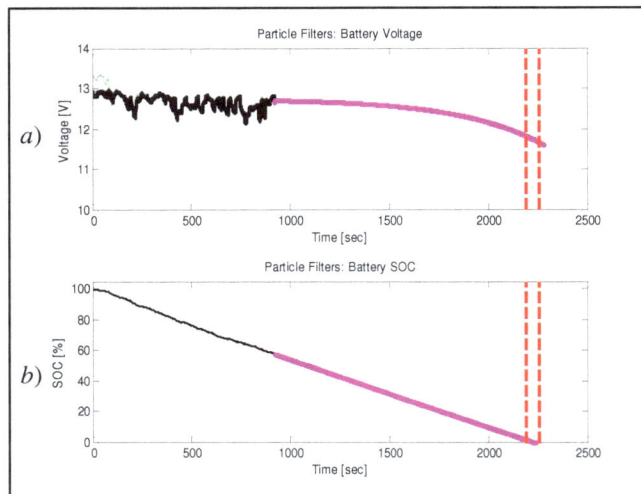

Figure 4: SOC prognosis assuming that the future discharge profile is a constant current equal to the average value measured in past data. (*a*) Measured voltage (thin green line), estimated voltage (dark black line), predicted voltage drop (magenta dashed line) and 95% confidence interval for EOD (dashed vertical lines). (*b*) Estimated SOC [%] (thin black line), predicted SOC [%] (magenta thick line), and 95% confidence interval for EOD (dashed vertical lines).

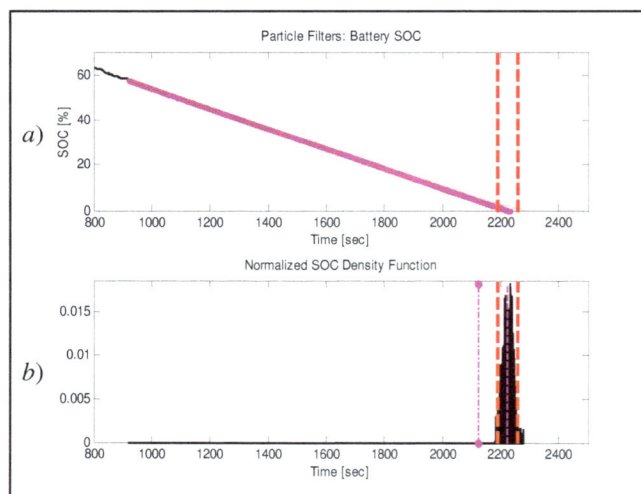

Figure 5: SOC prognosis assuming that the future discharge profile is a constant current equal to the average value measured in past data (zoom-in). (*a*) Estimated SOC [%] (thin black line), predicted SOC [%] (magenta thick line), 95% confidence interval for predicted SOC [%] (magenta thin lines)), and 95% confidence interval for EOD (dashed vertical lines). (*b*) Ground truth EOD (vertical magenta segment with markers), predicted EOD PDF, and 95% confidence interval for EOD.

A correct characterization of the tails of the PDF enable the implementation of much more sophisticated decision-making strategies, based on concepts such as the Just-in-

Time point (JITP) instead of the expectation of the distribution, with the purpose of avoiding failures in the system that could occur before the moment that is being predicted by the prognosis module.

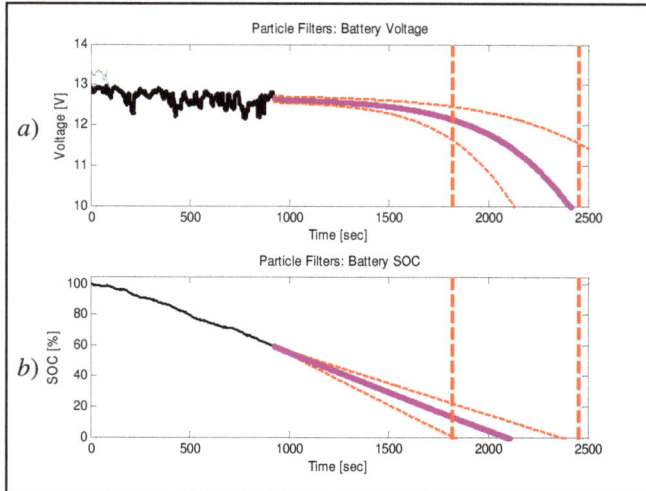

Figure 6: SOC prognosis assuming statistical characterization of future discharge profile. (*a*) Measured voltage (thin green line), estimated voltage (dark black line), predicted voltage drop (magenta dashed line) and 95% confidence interval for EOD (dashed vertical lines). (*b*) Estimated SOC [%] (thin black line), predicted SOC [%] (magenta thick line), and 95% confidence interval for EOD (dashed vertical lines).

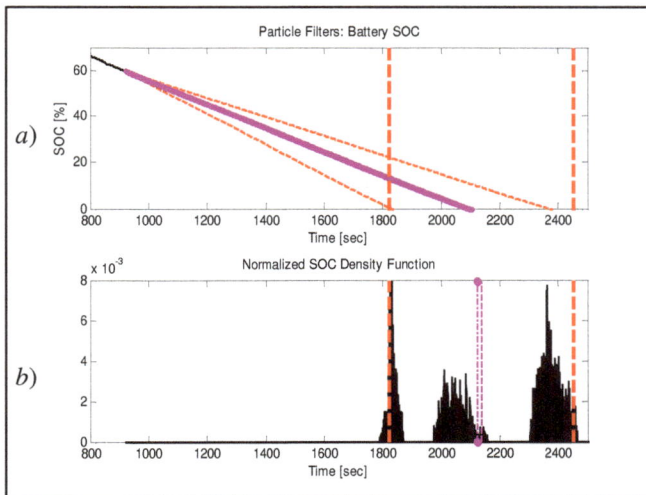

Figure 7: SOC prognosis assuming statistical characterization of future discharge profile. (zoom-in). (*a*) Estimated SOC [%] (thin black line), predicted SOC [%] (magenta thick line), 95% confidence interval for predicted SOC [%] (magenta thin lines)), and 95% confidence interval for EOD (dashed vertical lines). (*b*) Ground truth EOD (vertical magenta segment with markers), predicted EOD PDF, and 95% confidence interval for EOD.

One final remark can be made in term of the accuracy of the proposed algorithm if *ad-hoc* performance measures are to be used. In particular, Figure 8 shows the results obtained when using the "α-λ accuracy index" (Saxena *et al.*, 2010) ($\alpha = 15\%$; $\lambda = 0.5$). This measure determines if the predicted EOD is within a range defined by $\pm \alpha \%$ with respect to the true remaining time of operation, considering that the prediction window represents a fraction λ of the total time of operation. If the system were to be time-invariant, the remaining time should decrease linearly with slope equal to -1. Although this assumption does not necessarily characterize the true evolution of the autonomy of the system in this case, it still represents a good indicator on the consistency of the prognostic result. Figure 8 shows that the conditional expectation of the EOD consistently decreases as more data is acquired, a fact that is important to validate the proposed algorithm since Figure 6 and Figure 7 illustrate only the response for a unique prediction time (920 [sec]).

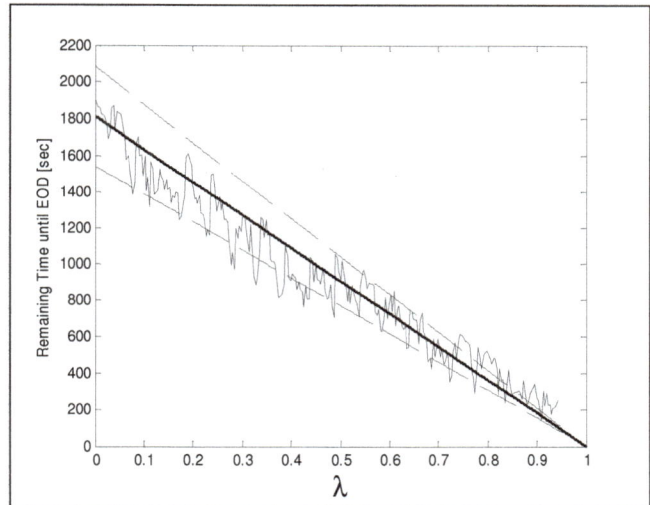

Figure 8: Prognosis performance evaluation based on the α-λ accuracy measure; α=15% and λ=0.5.

5. CONCLUSION

This paper presents the implementation of a particle-filtering-based prognostic framework that allows estimating the state-of-charge (SOC) and predicting the discharge time of energy storage devices, and more specifically lithium-ion batteries. The proposed approach uses an empirical state-space model inspired in the battery phenomenology and particle-filtering to study the evolution of the SOC in time; adapting the value of unknown model parameters during the filtering stage and enabling fast convergence for the state estimates that define the initial condition for the prognosis stage. SOC prognosis is implemented using a particle-filtering-based framework that considers uncertainty in the future discharge profile. The fact of incorporating more uncertainty in the long-term predictions clearly has an impact on the precision of the resulting 95% confidence interval; however, as a trade-off, the accuracy of the EOD expectation (2137.5 [sec]) is highly increased (see Figure 7) to a point where the algorithm has an error of only 13.8 [sec] in a 1203.7 [sec] prediction window. Although the precision of the proposed algorithm is much worse that in the case when constant discharge current is assumed, it characterize in a better manner all uncertainty sources and the tails of the PDF; enabling the implementation of much more sophisticated decision-making strategies, based on concepts such as the Just-in-Time point (JITP).

ACKNOWLEDGEMENT

The authors want to thank CONICYT for the financial support provided through project FONDECYT #1110070.

REFERENCES

Saha, B. and Goebel, K., (2009). "Modeling Li-ion Battery Capacity Depletion in a Particle Filtering Framework," Annual Conference of the Prognostics and Health Management Society, San Diego, CA.

Ranjbar, A.H., Banaei, A., Khoobroo, A., Fahimi, B., (2012). "Online Estimation of State of Charge in Li-Ion Batteries Using Impulse Response Concept," *Smart Grid, IEEE Transactions on* , Vol. 3, No.1, pp.360-367.

Pattipati, B., Sankavaram, C., and Pattipati, K., (2011). "System Identification and Estimation Framework for Pivotal Automotive Battery Management System Characteristics," *Systems, Man, and Cybernetics, Part C: Applications and Reviews, IEEE Transactions on*, Vol.41, No.6, pp.869-884.

Orchard, M., and Vachtsevanos, G., (2009). "A Particle Filtering Approach for On-Line Fault Diagnosis and Failure Prognosis," *Transactions of the Institute of Measurement and Control*, vol. 31, no. 3-4, pp. 221-246.

Salkind, A.J., Fennie, C., Singh, P., Atwater, T., Reisner, D.E., (1999). "Determination of state-of-charge and state-of-health of batteries by fuzzy logic methodology," *Journal of Power Sources*, Vol. 80, Issue 1-2, pp. 293-300.

Charkhgard, M., and Farrokhi, M., (2010). "State-of-Charge Estimation for Lithium-Ion Batteries Using Neural Networks and EKF," *Industrial Electronics, IEEE Transactions on*, Vol.57, No.12, pp.4178-4187.

Vinh Do, D., Forgez, C., El Kadri Benkara, K., Friedrich, G., (2009). "Impedance Observer for a Li-Ion Battery Using Kalman Filter," *Vehicular Technology, IEEE Transactions on*, Vol.58, No.8, pp. 3930-3937.

Ran, L., Junfeng, W., Haiying, W., Gechen, L., (2010). "Prediction of state of charge of Lithium-ion rechargeable battery with electrochemical impedance spectroscopy theory," Industrial Electronics and Applications (ICIEA), 2010 the 5th IEEE Conference on, Vol., No., pp.684-688.

Cadar, D.V., Petreus, D.M., Orian, C.A., (2009). "A method of determining a lithium-ion battery's state of charge," 15th International Symposium for Design and Technology of Electronics Packages (SIITME) 2009, pp.257-260.

Qingsheng, S., Chenghui, Z., Naxin, C., Xiaoping, Z., (2010). "Battery State-Of-Charge estimation in Electric Vehicle using Elman neural network method," 29th Chinese Control Conference (CCC) 2010, pp. 5999-6003.

Di Z., Yan, M., Qing-Wen, B., (2011). "Estimation of Lithium-ion battery state of charge," 30th Chinese Control Conference (CCC) 2011, pp. 6256-6260.

Tang, X., Mao, X., Lin, J., Koch, B., (2011). "Li-ion battery parameter estimation for state of charge," American Control Conference (ACC) 2011, Vol., No., pp. 941-946.

Saha, B., Goebel, K., Poll, S., Christophersen, J., (2009). "Prognostics Methods for Battery Health Monitoring Using a Bayesian Framework," *IEEE Transactions on Instrumentation and Measurement*, Vol.58, No.2, pp. 291-296.

Dalal, M., Ma, J., and He, D., (2011). "Lithium-ion battery life prognostic health management system using particle filtering framework," *Proceedings of the Institution of Mechanical Engineers, Part O: Journal of Risk and Reliability*, 225: 81-90.

Lee S., Kim J., Lee J., and Cho B.H., (2011). "Discrimination of Li-ion batteries based on Hamming network using discharging–charging voltage pattern recognition for improved state-of-charge estimation," *Journal of Power Sources*, v196, n4, pp. 2227-2240.

Santhanagopalan S., and White R.E., (2010). "State of charge estimation using an unscented filter for high power lithium ion cells," *International Journal of Energy Research*, v34, n2, pp. 152-163.

Hu C., Youn B.D., and Chung J., (2012). "A multiscale framework with extended Kalman filter for lithium-ion battery SOC and capacity estimation," *Applied Energy*, v92, pp. 694–704.

He Y., Liu X.T., Zhang C.B., Chen Z.H., (2013). "A new model for State-of-Charge (SOC) estimation for high-power Li-ion batteries", *Applied Energy*, v101, pp. 808-814.

Orchard, M., Tang, L., Saha, B., Goebel, K., and Vachtsevanos, G., (2010) "Risk-Sensitive Particle-Filtering-based Prognosis Framework for Estimation of Remaining Useful Life in Energy Storage Devices," *Studies in Informatics and Control*, Vol. 19, Issue 3, pp. 209-218.

Arulampalam, M., Maskell, S., Gordon, N., and Clapp, T., (2002). "A Tutorial on Particle Filters for Online Nonlinear/Non-Gaussian Bayesian Tracking,'" *IEEE Transactions on Signal Processing*, Vol. 50.

Andrieu C., Doucet A., Punskaya E., (2001). "Sequential Monte Carlo Methods for Optimal Filtering," in *Sequential Monte Carlo Methods in Practice*, A. Doucet, N. de Freitas, and N. Gordon, Eds. NY: Springer-Verlag.

Doucet A., de Freitas N., Gordon N., (2001). "An introduction to Sequential Monte Carlo methods," in *Sequential Monte Carlo Methods in Practice*, A. Doucet, N. de Freitas, and N. Gordon, Eds. NY: Springer-Verlag.

Engel, S.J., Gilmartin, B.J., Bongort, K., Hess, A., (2000). "Prognostics, the real issues involved with predicting life remaining," Aerospace Conference Proceedings, 2000 IEEE, Vol.6, pp.457-469.

Orchard, M., Tobar, F., Vachtsevanos, G., (2009) "Outer Feedback Correction Loops in Particle Filtering-based Prognostic Algorithms: Statistical Performance Comparison," *Studies in Informatics and Control*, vol. 18, Issue 4, pp. 295-304.

Edwards, D., Orchard, M., Tang, L., Goebel, K., Vachtsevanos, G., (2010) "Impact of Input Uncertainty on Failure Prognostic Algorithms: Extending the Remaining Useful Life of Nonlinear Systems," Annual Conference of the Prognostics and Health Management Society 2010, Portland, OR, USA.

Chen, C. Vachtsevanos, G., Orchard, M., (2011) "Machine Condition Prediction Based on Adaptive Neuro-Fuzzy and High-Order Particle Filtering," *IEEE Transactions on Industrial Electronics*, vol. 58, no. 9, pp. 4353-4364.

Zhang, B., Sconyers, C., Byington, C., Patrick, R., Orchard, M., and Vachtsevanos, G., (2011). "A Probabilistic Fault Detection Approach: Application to Bearing Fault Detection," *IEEE Transactions on Industrial Electronics*, Vol. 58, No. 5, pp. 2011-2018.

Saxena, A., Celaya, J., Saha, B., Saha, S., Goebel, K., (2010). "Evaluating prognostics performance for algorithms incorporating uncertainty estimates," *Aerospace Conference, 2010 IEEE*, pp.1-11.

Tang, L., Orchard, M.E., Goebel, K., Vachtsevanos, G., (2011). "Novel metrics and methodologies for the verification and validation of prognostic algorithms," Aerospace Conference, 2011 IEEE, pp.1-8.

Gonzalez G.D., Orchard, M., Cerda J.L., Casali A. and Vallebuona, G., (2003). "Local models for soft-sensors in a rougher flotation bank," Minerals Engineering, vol. 16, no.5, pp. 441-453.

Hunter, J. S., (1986). "The exponentially weighted moving average," *J. Qual. Technol.*, Vol. 18, pp. 203-209.

BIOGRAPHIES

Dr. Marcos E. Orchard is Assistant Professor with the Department of Electrical Engineering at Universidad de Chile and was part of the Intelligent Control Systems Laboratory at The Georgia Institute of Technology. His current research interest is the design, implementation and testing of real-time frameworks for fault diagnosis and failure prognosis, with applications to battery management systems, mining industry, and finance. His fields of expertise include statistical process monitoring, parametric/non-parametric modeling, and system identification. His research work at the Georgia Institute of Technology was the foundation of novel real-time fault diagnosis and failure prognosis approaches based on particle filtering algorithms. He received his Ph.D. and M.S. degrees from The Georgia Institute of Technology, Atlanta, GA, in 2005 and 2007, respectively. He received his B.S. degree (1999) and a Civil Industrial Engineering degree with Electrical Major (2001) from Catholic University of Chile. Dr. Orchard has published more than 50 papers in his areas of expertise.

M.S. Matías A. Cerda was born in Santiago, Chile in 1987. He received both his B.Sc. and M.S. degrees in Electrical Engineering (2011) from Universidad de Chile. He currently is a Research Engineer at the "Lithium Innovation Center" (Santiago, Chile). His research interests include prognostics and health management for energy storage devices based on optimal and suboptimal Bayesian algorithms.

M.S. Benjamín E. Olivares was born in Santiago, Chile in 1987. He received both his B.Sc. and M.S. degrees in Electrical Engineering (2011) from Universidad de Chile. He currently is a Research Engineer at the "Lithium Innovation Center" (Santiago, Chile). His research interests include prognostics and health management for energy storage devices based on optimal and suboptimal Bayesian algorithms.

Dr. Jorge F. Silva is Assistant Professor at the Department of Electrical Engineering, University of Chile, Santiago, Chile. He received the Master of Science (2005) and Ph.D. (2008) in Electrical Engineering from the University of Southern California (USC). He is IEEE member of the Signal Processing and Information Theory Societies and he has participated as a reviewer in various IEEE journals on Signal Processing. Jorge F. Silva is recipient of the Outstanding Thesis Award 2009 for Theoretical Research of the Viterbi School of Engineering, the Viterbi Doctoral Fellowship 2007-2008 and Simon Ramo Scholarship 2007-2008 at USC. His research interests include: non-parametric learning; sparse signal representations, statistical learning; universal source coding; sequential decision and estimation; distributive learning and sensor networks.

A Mobility Performance Assessment on Plug-in EV Battery

Seyed Mohammad Rezvanizanian, Yixiang Huang, Jiang Chuan, and Jay Lee

Center for Intelligent Maintenance System, University of Cincinnati, Cincinnati, Ohio, 45220, USA

rezvansd@mail.uc.edu
huangyx@mail.uc.edu
jiangcn@mail.uc.edu
lj2@mail.uc.edu

ABSTRACT

This paper deals with mobility prediction of LiFeMnPO$_4$ batteries for an emission-free Electric Vehicle. The data-driven model has been developed based on empirical data from two different road types –highway and local streets –and two different driving modes – aggressive and moderate. Battery State of Charge (SoC) can be predicted on any new roads based on the trained model by selecting the drving mode. In this paper, the performance of Adaptive Recurrent Neural Network (ARNN) and regression is evaluated using two benchmark data sets. The ARNN model at first estimates the speed profile of the new road based on slope and then both slope and speed is going to be used as the input to estimate battery current and SoC. Through comparison it is found that if ARNN system is appropriately trained, it performs with better accuracy than Regression in both two road types and driving modes. The results show that prediction SoC model follows the Columb-counting SoC according to the road slope. [1]

1. INTRODUCTION

Concerns with fuel cost, oil shortages, air pollution, and higher fuel economy standards have driven the rapid rise of more fuel-efficient vehicles and Electric Vehicles (EV). The future of transportation is being propelled by a fundamental move to green and more efficient electric drive systems. However, Electric vehicles still represent a small part of the worldwide market. For example, EVs account for just over 1% of the passenger car market at present (Shafiei & Williamson, 2010). With recent major advancements in battery technology, however, more electric vehicles are anticipated to enter the market within the next few years.

New concepts and technologies need to be developed to launch electrically chargeable vehicles suited for both individual and public mobility and for goods distribution in urban areas. Electrically charged vehicles provide many benefits in metropolitan areas, such as very low (plug-in hybrid electric vehicle - PHEV) to zero (battery electric vehicle - BEV) tailpipe emissions and reduced noise.

Although battery technology has developed considerably in recent decades, the main drawback of electric vehicles still remains: the low range of such vehicles in comparison to their gas-powered counterparts. Engineers and EV designers are challenged to improve EV mobility performance with existing battery power. The lead-Acid battery - and its variants- is the dominant battery in the automotive industry due to its low manufacturing price, high C-rate discharge and good low temperature performance. However, the low energy density of these batteries makes them unsuitable for hybrid and electric vehicle applications. Currently, car manufacturers are motivated to utilize Ni-MH and Li-ion batteries because their high energy density and cycle life that satisfy the requirements of an electric vehicle. Therefore accurate estimation of the SoC of a high capacity energy storage system can improve energy management and efficient utilization of electric vehicle by optimizing performance, lengthening the cycle life and providing more useful information for the driver.

The output power from the battery depends on a number of factors, such as discharge current rate, internal resistance, battery age, environment temperature and historical usage (Meissner & Richter, 2003; Ulrich, 2012). All above factors influence SoC in linear and non-linear ways. Therefore, a number of diverse techniques have been proposed to calculate or estimate the SoC of a battery each of which as its relative advantages and constraints, as reviewed by (Piller, Perrin, & Jossen, 2001; Rodrigues, Munichandraiah, & Shukla, 2000; Zhang & Lee, 2011). However, all of these techniques attempt to monitor and measure the SoC at the current time. Applying prognostics algorithms can help engineers predict battery conditions as well. Fully informed travelers and increased productivity of energy storage devices are the requirements for enabling the intelligent mobility of an electric vehicle. Due to the broad availability of low cost wireless communication systems, related information flows constantly and flawlessly from each source to all interested users. Using these systems provides an opportunity for drivers to obtain directions and optimize their route before starting a trip. Moreover, these devices can provide the raw data necessary for prognostics algorithms to predict the conditions of the battery as well.

This paper investigates new methods of using machine learning techniques and prognostics algorithms to predict the battery SoC of an EV prior to starting a trip, which is based on the route selected and historical driving behavior by the driver. Section 2 gives an overview of why the market is looking for SoC estimation and EV battery performance prediction. The details of empirical testing on EV are presented in section 3. Section 4 provides the methodology and details of algorithms used for prediction. The details of the data analysis, including feature extraction and SoC estimation, are presented in section 5. Finally, the results and conclusion are discussed in section 6 and 7.

2. PROBLEM STATEMENT

One of the most important barriers in acceptance of electric mobility is a range of electric vehicle. It has been evaluate the effects of low range resources of electric vehicles, as a significant feature for users' purchase intentions by market experts and prospective customers. As it has been mentioned before new communication technology provides much more data to the interested user. A GPS can provide some basic route options including local streets, highways or combination of both with an estimation of travel time. However, such estimations are not tailored for the specific needs of EV users. There is no prediction of energy consumption for suggested routes, or any inclusion of

Figure 1. Two different types of road (highway and local streets)

driving behavior or road conditions. Route suggestions based on this information would go a long way towards alleviating potential range anxiety that is one of the major barriers for EV adoption.

In battery prognostics, since it is a relatively new field of interest, there are few works that can fit inside the requirements. For example, (Gonder, Markel, Thornton, & Simpson, 2007) found that around 95% of daily driving can be achieved with 100 miles of electric range by studying a set of vehicles for thirty weekdays each. Correspondingly, studies by (MacLean & Lave, 2003) and (Sioshansi & Denholm, 2009) on hybrid vehicles have determined that about three-quarters of travel miles could be powered by electricity. The main issue with such studies is that customers desire a vehicle that fulfills their own diverse needs over time, not the needs that are dictated by statistics or the average profile. In general, changing the charge status of the battery is a complex process, resulting from the interaction of several factors. Among these factors, discharging current rate is the most significant, and it depends on several car resistances. For mechanical power generation of a vehicle, the electric motor should be able to provide enough power to accelerate and propel the EV if it encounters any resistance. There are three significant forms of resistance for a vehicle: tire friction, aerodynamic resistance and gravity resistance (Fodor, Enisz, Doman, & Toth, 2011; Khaled, Harambat, Yammine, & Peerhossaini, 2010; Shukla, Aricò, & Antonucci, 2001). These forms of resistance are directly influenced by vehicle speed and road slope. Moreover the driver's driving habits affected battery performance. Even the same maximum speed on a road with different accelerations can have different effects on the charge of battery. In this paper this term is called "driving mode" and is defined it as acceleration and braking. Hence this paper looks for a method to predict SOC based on three factors:

1. Vehicle Speed
2. Road Slope (Terrain type)
3. Driving Mode

Researchers have established different techniques for formulating a relationship between SoC and these factors. In (He, Xiong, Zhang, Sun, & Fan, 2011) an improved Thevenin model and utilized Extended Kalman Filter were used to calculate SoC for an EV which was driving at variable speeds. In (He, Xiong, & Guo, 2012; Xu, Wang, & Chen, 2012) the battery was modeled with combination of Extended Kalman Filters and Fuzzy Neural Network to give an approximation of SoC. Some researchers attempted to determine realistic driving conditions and their effects on battery performance. In (Adornato, Patil, Filipi, Baraket, & Gordon, 2009; Lee, Baraket, Gordon, & Filipi, 2011) the authors classified practical driving to identify a charging model for each category of driving mode. In (Marina de Queiroz Tavares, 2010) GPS tracker is using to monitor where the EV is charging and where it goes in order to recognize driving behavior. What's missing is how to define the relationship between these factors and mobility.

In fact, the inclusion of all these variables would complicate the findings of SoC prediction. For clarification purposes, this paper will focus on one variable and its effect on battery performance, energy efficiency, and long-term operation on battery Remaining Useful Life (RUL). Hence, several tests have been designed on the EV to be tested on different roads with different slopes and see how the slope can affect battery performance.

3. EXPERIMENTAL DESCRIPTION

The best approach to represent real–world driving patterns is to test an EV on specific routes and collect voltage and current from battery and speed from car. A major drawback of this approach is that it is difficult to justify the value that can be derived from such costly experiments. Because of the sporadic nature of driving cycles and the wide distribution of driving conditions, obtaining meaningful information from field testing for accurate analysis is very difficult. Researchers previously summarized statistical results from field tests to offer some help in understanding battery performance (Huang, Tan, & He, 2011; Lee et al., 2011; Liaw & Dubarry, 2007; Montazeri-Gh, Fotouhi, & Naderpour, 2011). According to the literature, road type, traffic road slope, traffic and driving mode have major affects on battery state of charge and energy consumption during a trip; however, there are some other factors such as ambient temperature, humidity, and charging intervals that can affect on battery life on long term operation. Since considering all factors make the problem very complex and difficult to solve, in this paper we set these factors to constant values and conduct the test on just different road types and with driving modes to investigate battery performance in diverse conditions. Data has been collected from a Chevy Equinox (Figure 2), which has been converted from a regular gas powered car to an emissions-free Electric Vehicle by AMP. This car has been equipped with LiFeMnPO$_4$ battery from the GBSystem. The basic specifications of the EV are summarized in Table 1.

Characteristics	Range	Unit
Car Range	90-100	Miles
Max Speed	90	mph
Number of Cells	108	
Cell connection	in-series	
Total Battery Energy	40	kWh
Battery type	LiFeMnPO$_4$	
Nominal Capacity	100	A.h

Table 1. Chevy Equinox and battery pack specification

To make the condition of the tests as stable as possible all tests have been done using a single car. In each test two persons were in the car (the driver and one passenger were to control data collection). The condition of test is based on two road types: highways and local streets. Both road types were selected to have some uphill and downhill conditions to observe battery performance in both conditions.

Figure 2. Chevy Equinox electric vehicle

4. METHODOLOGY

One of the most proficient ways of solving a multifaceted problem is to decompose it into simpler elements, in order to make it more understandable and more manageable (Bo, Zhifeng, & Binggang, 2008). In addition, simple elements may be assembled to produce a complex system. One approach for achieving this is using Networks. All networks consist of nodes and connections, where the nodes can be considered to be computational units. Nodes receive inputs and apply some mathematical processes to attain an output. This processing might be very straightforward (such as summing the inputs), or quite complex.

The Neural Network (NN) techniques are one of the common methods, which typically consist of inputs that are multiplied by weights (strength of the particular signals), and then computed by a mathematical function that represents the activation of the neuron. Another function computes the output of the artificial neuron (sometimes in dependence of a certain threshold). NNs combine artificial neurons in order to find relations among inputs and outputs (Charkhgard & Farrokhi, 2010). In this paper Adaptive Recurrent Neural Network is implemented on EV data. Recurrent or recursive networks are well suited to time variant modeling applications, such as prediction. This is because temporal knowledge is saved by the network in the form of time-delayed inputs; outputs from one iteration of the model are fed back as inputs into one or more succeeding iterations (Wang, Golnaraghi, & Ismail, 2004).

In this paper the recurrent network has a feedback link from outputs to the inputs, which serves as the third input layer. This layer is illustrated in Figure 3. However, in some cases this recurrent network can be applied from the hidden layer to the context (Wang et al., 2004). This network has three input nodes (slope, speed and previous step of current), three layers and one output. The nodes in the input simply send out input values to the hidden layer. For the nodes in the hidden layers, hyperbolic tangent sigmoid functions have been selected as their transfer functions, while a linear function is assigned to the output node. By training the model, the connection weights can be adjusted in the algorithm. The higher the weight of an artificial neuron the stronger the input is; meaning that this specific input is more significant (Abolhassani Monfared et al., 2006; Julka et al., 2011). Eq. (1) shows the function of output I at time k of three inputs where **p** is the slope of the road; **s** is the speed of the car; and **I** is the current at one step ago.

$$I(k) = f(P(k-1), S(k-1), I(k-1)) \qquad (1)$$

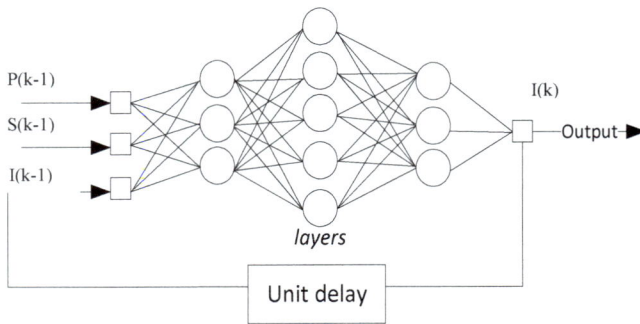

Figure 3. Adaptive Recurrent Neural Network Concepts

A Linear Regression method has been applied on the data to benchmark the accuracy of results with ARNN. In the regression process, a linear model has been built between the inputs and outputs by applying training data. The model is then tested by new inputs and tested with the measured output to calculate error.

5. DATA ANALYSIS AND DISCUSSION

5.1. Feature extraction

In addition to the aforementioned methods, it is necessary to find the operational parameters that change with different driving modes in order. The accuracy of the battery performance estimation and SoC prediction will heavily rely on these so-called features. From the raw data, many features could be extracted, but not all the features will be directly related to the driving behavior. In this paper, driving mode has been classified into two categories: aggressive and moderate. Currently, two main features are extracted from battery current. To have better understanding of the two different driving modes, Figure 4 shows battery current for aggressive and moderate driving modes on the same local streets with the same driver and the same car.

Figure 5 a) and 5 b) illustrate the distribution of current for these two different driving modes on one type of road (local streets). The aggressive driving mode is distributed in a lager range of current: the minimum value is around -120 A and the maximum around 200 A. The moderate driving mode has changed, however, to between -85 A to 140 A.

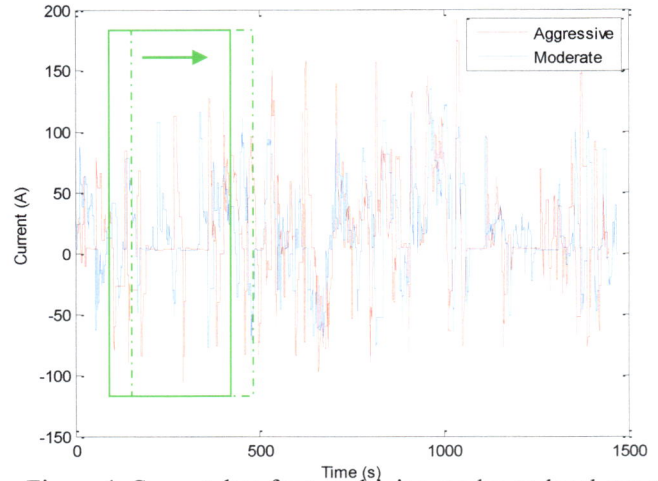

Figure 4. Current data for two driving modes on local streets

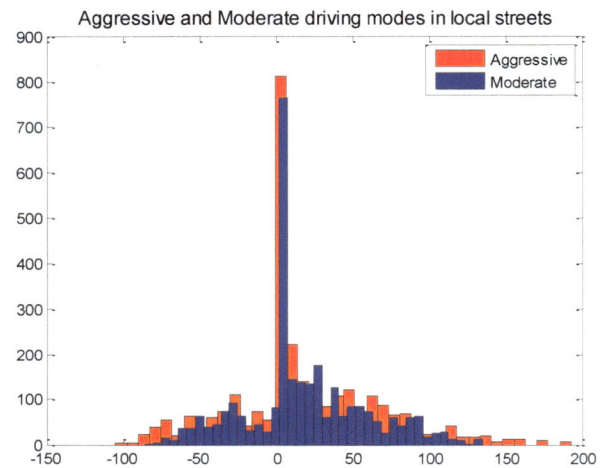

Figure 5 a). Histogram for local streets

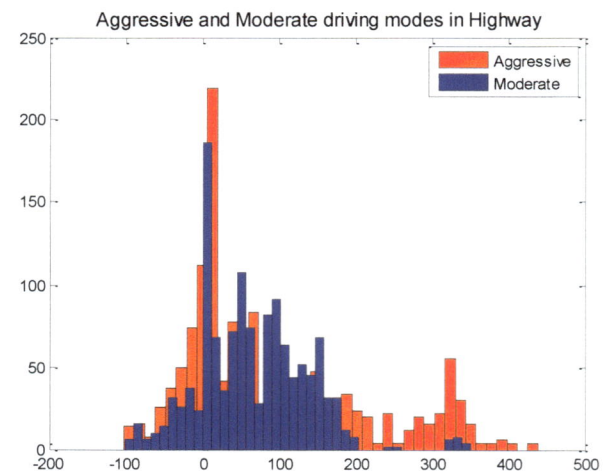

Figure 5 b). Histogram for highway

For highway mode this range has been extended due o the higher speeds reached, since the aggressive driver reaches more than 400 A. The other significant difference is related to the high frequency of near zero value for current data. In fact, any time the driver increases speed from zero and implements hard braking, a value close to zero is recorded. This is inevitable when the car is approaching traffic lights, which is a pronounced difference between local streets and highways. Highways do not often contain stop points, that can help us to recognize the route type based on battery current data analysis.

To obtain a mathematical threshold that can help us distinguish between the two driving modes, the standard deviation of both currents has been calculated in the specific window size, by moving the window forward in each by one step in value of standard deviation recorded (Figure 4). Figure 6 represents the results of how the standard deviation for current changes over time. The average of aggressive driving mode standard deviation is 44.3 and the moderate one is 35.6.

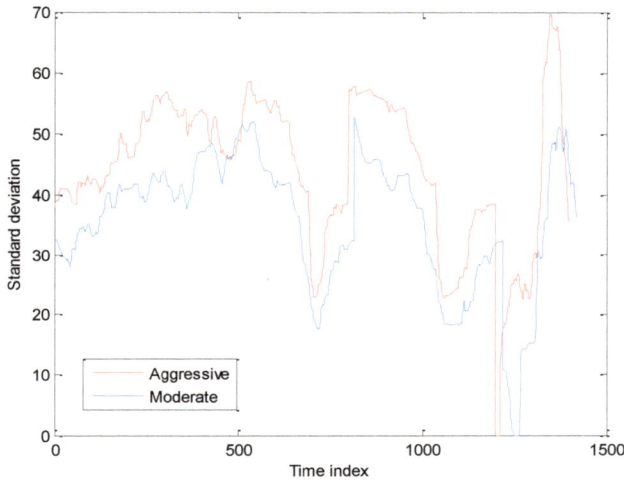

Figure 6. Battery current standard deviation over time

5.2. SOC Estimation

The estimation of the SoC of a battery may be a problem of more or less complexity depending on the battery type and on the application in which the battery is used. The most reliable technique to define the SoC for Electric Vehicles during the operation is ampere hour counting (Columb counting) (Zhu, Coleman, & Hurley, 2004). The idea of balancing the current is reasonable because the charging and discharging cycles are directly related to the supplied or demanded current. If the initial SoC value is given (SoC_0), the value of current integral is a straight indicator for the SoC. Eqs. 2 and 3 show this relation for both charge and discharge respectively:

$$SoC = SoC_0 + \frac{1}{C_n} \cdot \int_{t_0}^{t} |I| \cdot dt \qquad charge \qquad (2)$$

$$SoC = SoC_0 - \frac{1}{C_n} \cdot \int_{t_0}^{t} |I| \cdot dt \qquad discharge \qquad (3)$$

Where C_n is the nominal capacity of the EV battery pack, I is the battery current and dt is the time interval. It is obvious that SoC is a function of current. Therefore if we can estimate the current of the vehicle before it completes its route then it is possible to calculate SoC. For this purpose we need to figure out how that current is changing based on road condition.

Based on these assumptions, the amount of current demanded from a battery dependent on how fast the car moves, or accelerates and the slope of the road. The main challenge is finding what the relation is between road slope and car speed. Theoretically, the car speed can be independent from the slope. However, in real-world data, we can see that by driving up a hill the speed can be diminished. Actually, there is no definite answer for this question because controlling the speed depends on the driver's decision to keep the speed constant, let it reduce a little or even accelerate. In this paper the car has been assumed that speed is independent from the road slope.

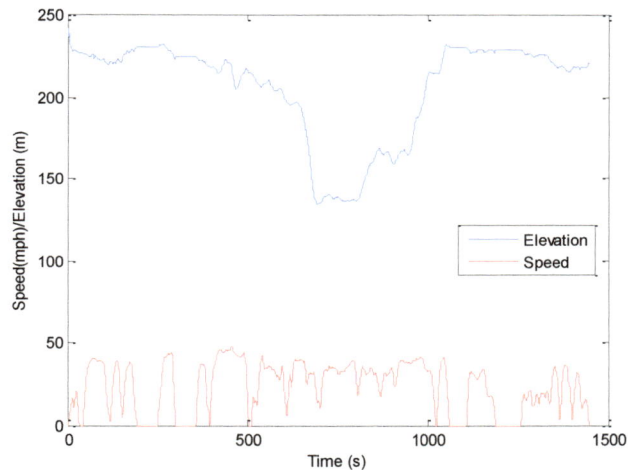

Figure 7. Elevation and Speed in a local street; there are input variables in the model

The ARNN predictor has been constructed based on a feed-forward multi-layer neural network with adaptive and feedback links from input nodes. The feedback units reproduce the activations of the nodes from the previous time step, and allow the network to memorize the evidence from the past, which forms a reasoning base for current processing. For the nodes in the hidden layers, the sigmoid activation function has been used and the linear activation function has been selected for the nodes in the output layer. The strategy is to train the model based on the ARNN with speed and slope at the existing time step and current as the output of the model from the previous step as the input variables for two different driving modes (Figure 7) with current as the target (Figure 4 red color). Figure 7 represents the sample input for training, which is road elevation and speed for a local street under an aggressive driving mode together. Table 2 summarizes the process of constructing the training data set.

	Road type	Driving mode	Length of the road
1	Highway 1	Aggressive	10 miles
2	Highway 1	Moderate	10 miles
3	Local St 1	Aggressive	10 miles
4	Local St 1	Moderate	10 miles

Table 2. Condition of training dataset

In addition to this aforementioned challenge, for testing the model, the car velocity on the road is needed. It is not possible to obtain the exact velocity profile before performing the whole route. Since aggressiveness can affect the speed of the car as well, we have used the same approach to obtain the speed. The ARNN model has been trained with just slope data as input and speed as the target variable. Again one step of the speed can be used as the input for the model for the step k+1. This training has been done for all four combinations of driving modes and road types. Figure 8 illustrates the result of this type of estimation of the speed profile for two different road types in moderate driving modes.

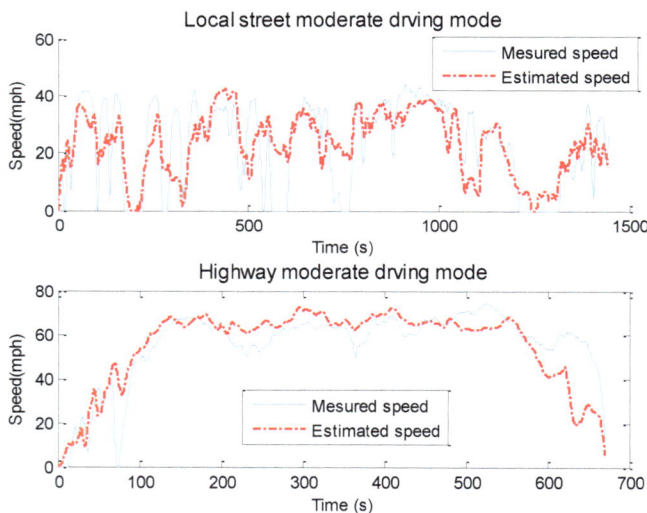

Figure 8. Estimated speed; testing for highway and local street roads

Looking at the speed predictions for the local street route shows that whenever the actual speed is zero the model cannot predict the speed accurately. The reason is that, since the model is trained based on data from another type of road, it does not have enough information from the new road except slope. Even though slope can provide some information for acceleration or deceleration of speed, it cannot give any information when the car is stopping. Some

other parameters, such as the location of traffic lights and stop signs, can affect this model. But, for highway this issue does not appear and the model can provide a better estimation of the speed. The output of this model (speed) is going to be used as the input for the SoC prediction in next step.

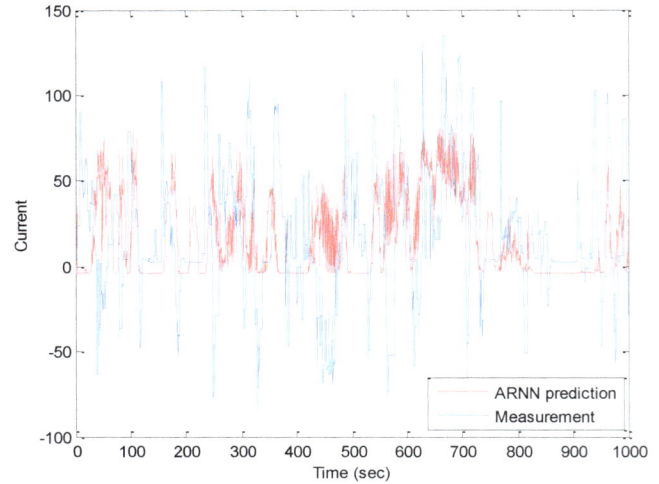

Figure 9. Comparing measured current and predictive current

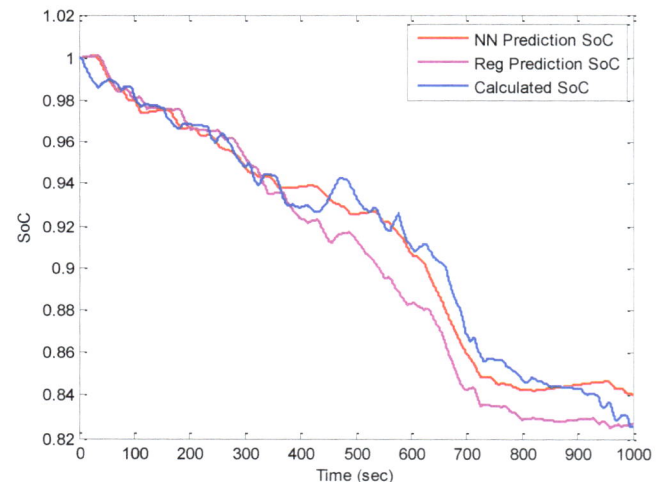

Figure 10. Comparing two methods of SoC estimation with Coloumb counting (measured)

Four different models have been trained and has been represented in Table 2. These models are the combination of each road type (local streets and highway) with one of driving modes (aggressive and moderate), so with new road data (slope of the road) the model should be able to predict the battery current profile. During testing the appropriate model will be chosen based on the types of data be it highway or local in terms of road type, and aggressive or moderate for driving behavior. Figure 9 shows the results of battery current prediction based on the second road slope and the estimated speed from the previous step with moderate driving mode. This estimation has been done with the ARNN model. The same approach has also been done with linear regression analysis. According to Eq. (3) the SoC can be calculated, which has been represented on Figure 10.

6. RESULTS AND DISSCUSSION

Figure 9 shows the behavior of the current and also shows that the prediction of the current follows the actual measurements. However, it is difficult to evaluate the accuracy of the model from the graph. Figure 10 shows that the prediction of SoC based on ARNN is accurate at certain points when compared to actual measurements, but the last value is not exactly correct. The linear regression gives a better estimation at the last point. To be able to evaluate both approaches the root mean square error has been calculated for four different testing data sets. The results have been summarized in Table 3.

	Road type	Driving mode	RMSE ARNN	RMSE Regression
1	Highway 2	Aggressive	12.87	23.32
2	Highway 2	Moderate	10.29	21.21
3	Local St 2	Aggressive	14.32	25.08
4	Local St 2	Moderate	11.98	22.11

Table 3. Benchmarking tests of RMSE in different road and driving condition

Comparing the results we can see that for both highway and local driving aggressive driving behavior scenarios exhibit a higher error rate in prediction. If we remember the feature extraction from driving modes, the standard deviation of the aggressive driving mode is higher which indicates that the current fluctuates much more compared to the moderate driving mode. And if the model cannot follow all of the fluctuations in current, the error increases. Another result shows that local streets have much more error in terms of prediction than Highways in the same driving modes. Based on empirical data, EVs consume a higher amount of energy in highway driving because of the higher average, though stable, speed, however, in local driving the average speed is much lower but it changes often due to traffic and stoplights, among other reasons.

Benchmarking the two algorithms (ARNN and regression) identifies accurate results from ARNN. The strength of ARNN is that they provide a dynamic modeling of the current and SoC. Figure 10 shows that SOC prediction based on ARNN follows the real values with a smaller error. However, we need to consider that this prediction has been applied on a small set of data. Larger data sets can help to enable much more accurate results.

7. CONCLUSION

Estimating battery state of charge is one of the most significant issues for electric vehicles. Since there are many factors that can affect battery SoC during vehicle operation, it is not easy to assess battery charge status. Discharging current rate is one of the factors that change dynamically during battery operation based on road slope, and car speed. Even if slope of the road is constant, different drivers can have disparate driving styles on the same road. In this paper a simple classification has been done on both driving modes and road types, showing how these two elements impact the current of the vehicle.

Whereas the GPS data is available in a car the GPS can show more information to the driver before in a trip. The objective is to utilize this data as the input and give the driver accurate SoC estimation, based on the different routes that the GPS provides. Two techniques have been applied in this paper. The Adaptive Recurrent Neural Network result is accurate in comparison to measured data and has a lower error rate. However, even the ARNN results still contain some error, which may be attributed to assumptions.

In fact, the method to predict the velocity profile is one of the reasons for the error. There are a variety of road condition factors that are not incorporated in this model, such as street junctions, stop signs, and traffic jams. Considering all these items in future plans can improve the speed prediction. Regardless of the velocity, this method relies on training data; the performed models are based on one road from each type. In this experiment just two roads of each type were selected. In reality there are an extremely large number of roads and road conditions. In the future it would be possible to upload data from all EVs in a cloud based system during their performance, and then apply machine learning tools easily cluster them in different regimes like different driving modes. It would then be easier to build a data driven model for prediction SoC.

ACKNOWLEDGEMENT

We would like to thank AMP Electric Vehicle Company and Mr. Don Wires for their assistance in providing infrastructure to collecting the vehicle test data.

REFERENCES

Abolhassani Monfared, N., Gharib, N., Moqtaderi, H., Hejabi, M., Amiri, M., Torabi, F., & Mosahebi, A. (2006). Prediction of state-of-charge effects on lead-acid battery characteristics using neural network parameter modifier. *Journal of Power Sources, 158*(2 SPEC. ISS.), 932-935.

Adornato, B., Patil, R., Filipi, Z., Baraket, Z., & Gordon, T. (2009). *Characterizing naturalistic driving patterns for plugin hybrid electric vehicle analysis.*

Bo, C., Zhifeng, B., & Binggang, C. (2008). State of charge estimation based on evolutionary neural network. *Energy Conversion and Management, 49*(10), 2788-2794.

Charkhgard, M., & Farrokhi, M. (2010). State-of-charge estimation for lithium-ion batteries using neural networks and EKF. *IEEE Transactions on Industrial Electronics, 57*(12), 4178-4187.

Fodor, D., Enisz, K., Doman, R., & Toth, P. (2011). *Tire road friction coefficient estimation methods comparison based on different vehicle dynamics models.*

Gonder, J., Markel, T., Thornton, M., & Simpson, A. (2007) Using global positioning system travel data to assess real-world energy use of plug-in hybrid electric vehicles. (pp. 26-32).

He, H., Xiong, R., & Guo, H. (2012). Online estimation of model parameters and state-of-charge of LiFePO4 batteries in electric vehicles. *Applied Energy, 89*(1), 413-420.

He, H., Xiong, R., Zhang, X., Sun, F., & Fan, J. (2011). State-of-charge estimation of the lithium-ion battery using an adaptive extended Kalman filter based on an improved Thevenin model. *IEEE Transactions on Vehicular Technology, 60*(4), 1461-1469.

Huang, X., Tan, Y., & He, X. (2011). An intelligent multifeature statistical approach for the discrimination of driving conditions of a hybrid electric vehicle. *IEEE Transactions on Intelligent Transportation Systems, 12*(2), 453-465.

Julka, N., Thirunavukkarasu, A., Lendermann, P., Gan, B. P., Schirrmann, A., Fromm, H., & Wong, E. (2011). Making use of prognostics health management information for aerospace spare components logistics network optimisation. *Computers in Industry, 62*(6), 613-622.

Khaled, M., Harambat, F., Yammine, A., & Peerhossaini, H. (2010). *Aerodynamic forces on a simplified car body - Towards innovative designs for car drag reduction.*

Lee, T. K., Baraket, Z., Gordon, T., & Filipi, Z. (2011). Characterizing One-day Missions of PHEVs Based on Representative Synthetic Driving Cycles. *SAE International Journal of Engines, 4*(1), 1088-1101.

Liaw, B. Y., & Dubarry, M. (2007). From driving cycle analysis to understanding battery performance in real-life electric hybrid vehicle operation. *Journal of Power Sources, 174*(1), 76-88.

MacLean, H. L., & Lave, L. B. (2003). Life Cycle Assessment of Automobile/Fuel Options. *Environmental Science and Technology, 37*(23), 5445-5452.

Marina de Queiroz Tavares, J. G., Flavio Perucchi, Franz Baumgartner, Maria Youssefzadeh. (2010). *Understanding future customer needs by monitoring EV-drivers' behavior.* Paper presented at the Hybrid and Fuel Cell Electric Vehicle Symposium & Exhibition, Shenzhen, China.

Meissner, E., & Richter, G. (2003). Battery Monitoring and Electrical Energy Management precondition for future vehicle electric power systems. *Journal of Power Sources, 116*(1-2), 79-98.

Montazeri-Gh, M., Fotouhi, A., & Naderpour, A. (2011). Driving patterns clustering based on driving feature analysis. *Proceedings of the Institution of Mechanical Engineers, Part C: Journal of Mechanical Engineering Science, 225*(6), 1301-1317.

Piller, S., Perrin, M., & Jossen, A. (2001). Methods for state-of-charge determination and their applications. *Journal of Power Sources, 96*(1), 113-120.

Rodrigues, S., Munichandraiah, N., & Shukla, A. K. (2000). Review of state-of-charge indication of batteries by means of a.c. impedance measurements. *Journal of Power Sources, 87*(1), 12-20.

Shafiei, A., & Williamson, S. S. (2010). *Plug-in hybrid electric vehicle charging: Current issues and future challenges.*

Shukla, A. K., Aricò, A. S., & Antonucci, V. (2001). An appraisal of electric automobile power sources. *Renewable and Sustainable Energy Reviews, 5*(2), 137-155.

Sioshansi, R., & Denholm, P. (2009). Emissions impacts and benefits of plug-in hybrid electric vehicles and vehicle-to-grid services. *Environmental Science and Technology, 43*(4), 1199-1204.

Ulrich, L. (2012). State of charge. *IEEE Spectrum, 49*(1), 56-59.

Wang, W. Q., Golnaraghi, M. F., & Ismail, F. (2004). Prognosis of machine health condition using neuro-fuzzy systems. *Mechanical Systems and Signal Processing, 18*(4), 813-831.

Xu, L., Wang, J., & Chen, Q. (2012). Kalman filtering state of charge estimation for battery management system based on a stochastic fuzzy neural network battery model. *Energy Conversion and Management, 53*(1), 33-39.

Zhang, J., & Lee, J. (2011). A review on prognostics and health monitoring of Li-ion battery. *Journal of Power Sources, 196*(15), 6007-6014.

Zhu, C. B., Coleman, M., & Hurley, W. G. (2004). *State of charge determination in a lead-acid battery: Combined EMF estimation and Ah-balance approach.*

An Intelligent Fleet Condition-Based Maintenance Decision Making Method Based on Multi-Agent

Qiang Feng[1], Songjie Li[2], and Bo Sun[3]

[1,2,3]*School of Reliability and Systems Engineering, Beihang University, Beijing, 100191, China*

fengqiang@buaa.edu.cn
songjieli@dse.buaa.edu.cn
sunbo@buaa.edu.cn

ABSTRACT

According to the demand for condition-based maintenance online decision making among a mission oriented fleet, an intelligent maintenance decision making method based on Multi-agent and heuristic rules is proposed. The process of condition-based maintenance within an aircraft fleet (each containing one or more Line Replaceable Modules) based on multiple maintenance thresholds is analyzed. Then the process is abstracted into a Multi-Agent Model, a 2-layer model structure containing host negotiation and independent negotiation is established, and the heuristic rules applied to global and local maintenance decision making is proposed. Based on Contract Net Protocol and the heuristic rules, the maintenance decision making algorithm is put forward. Finally, a fleet consisting of 10 aircrafts on a 3-wave continuous mission is illustrated to verify this method. Simulation results indicate that this method can improve the availability of the fleet, meet mission demands, rationalize the utilization of support resources and provide support for online maintenance decision making among a mission oriented fleet.

1. INTRODUCTION

When conducting a mission, an aircraft fleet consumes massive resources, especially maintenance manpower and resources. In practice, maintenance strategies usually combine the "fail and fix maintenance" with fixed preventive maintenance. The "fail and fix" strategy cannot prevent fatal accidents, which may endanger pilots' lives and reduce the mission availability, while fixed preventive maintenance strategy usually schedules excess maintenance actions to ensure availability, while ignoring the asynchronism of failures among a fleet and the shareability of maintenance resources, hence, cannot fully develop the

overall efficiency of maintenance resources, causing a huge waste while cannot completely prevent failure (Jiang & Murthy, 2008). Besides, to ensure safety, a specific maintenance job is done at a specific site, which may lead to the incoordination between operational requirements and maintenance actions. In general, traditional "fail and fix" practice & fixed preventive maintenance practice are not completely suitable.

To tackle the difficult problem, Condition-Based Maintenance (*CBM*), which is based on the actual condition and development tendency of assets, is put forward (Bengtsson, 2004). The rapid development of Prognostics and Health Management (*PHM*) (Sun, Zeng, Kang & Pecht, 2012) approach and its application on battery (Goebel, Saha, Saxena, Celaya & Christophersen, 2008) and aero engine (Wen & Liu, 2011) makes *CBM* possible. In practice, an aircraft contains one or more Line Replaceable Modules (*LRM*) whose health condition development fit the deterioration process (Barata, Guedes, Marseguerra & Zio., 2002). *PHM* can help predict the Residual Useful Life (*RUL*) of deteriorating *LRMs* through condition monitoring, and help staff make maintenance decision. Through the application of *PHM*, a series of maintenance measures are provided in time, and the ideal *CBM* "need and fix" is achieved (Jardine, Lin & Banjevic, 2006). Moreover, since *RUL* can be estimated, maintenance actions can be performed dynamically according to operational requirements rather than in a fixed site. In a fleet, where maintenance tasks are heavy and resources are limited, the application of *CBM* can notably increase operational availability, reduce lifecycle costs and improve safety.

Traditional *CBM* is about safely extending maintenance intervals using *PHM* information, and is often applied to a single aircraft. Fleet oriented *CBM*, on the other hand, should consider many factors other than single aircraft *CBM*, such as mission requirement, maintenance teams, etc., to balance the whole fleet. Actually, the ideal process of fleet *CBM* is as follows: 1) Aircrafts obtain their *PHM* data. 2) The *PHM* data is transferred to the maintenance center. 3)

The maintenance center makes maintenance decisions. 4) The maintenance decisions are transferred to aircrafts and maintenance teams. 5) Maintenance action. So the fleet *CBM* problem is actually an "online" decision making problem. Besides, the fleet maintenance strategy is the combination of maintenance strategies for every single aircraft. For each single aircraft, the problem is to find the most suitable time and team while balancing the whole fleet, which is actually a routing problem. Routing problem has already proved to be N-P hard (Garey & Johnson. 1979), which is difficult to obtain the optimal or satisfying solution with the increase of problem scale. At present, the main solutions to fleet *CBM* problem include

1. Mathematical programming: Doganay and Bohlin (2010) studied the train fleet maintenance scheduling strategy & spare parts optimization with single station based on a mixed integer linear programming. Bai (2009) optimized the Life Limited Part (LLP) group maintenance schedule and the on-wing lifetime of an aero engine fleet based on immune particle swarm method.

2. Heuristic method: Reimann, Kacprzynski, Cabral and Marini (2009) designed a maintenance scheduling algorithm combining *CBM* with traditional fixed preventive maintenance using heuristic method, to reduce the maintenance cost of a fleet consisting of 50 aircrafts, and to predict the shortage of maintenance resources.

3. System simulation: Bivona and Montemaggiore (2005) tested different maintenance & management strategies based on system dynamics modeling and simulation. Dupuy, Wesely and Jenkins (2011) selected the best one out of three civil aviation fleet maintenance plans applying discrete event simulation with the help of ARENA®.

4. Artificial Intelligence: Cycon (2011) discussed the technique Sikorsky Aircraft Corporation (SAC) is applying to incorporate *CBM* capabilities into all its products. By data collecting from all products and data mining, normal versus anomalous behavior is established, and man-in-the-loop support allows experts from various engineering and support services groups to quickly recommend appropriate maintenance actions. Zhou, Fox, Lee and Nee (2004) applied Multi-Agent technique and heuristic rules to solve the bus maintenance scheduling problem, which has equal optimality to reported studies and requires less computing time.

5. Multiple criteria analysis: Papakostas, Papachatzakis, Xanthakis, Mourtzis and Chryssolouris (2010) applied the multiple criteria (Cost, *RUL*, Operational Risk & Flight Delay) based on specific mission to select the best out of a set of generated maintenance plan alternatives using Monte-Carlo simulation.

But there are still shortages between those methods and dynamic environments where online maintenance decision making and scheduling is required when an aircraft fleet execute combat tasks. Especially in:

1. Those methods lack consideration into the relationship between the health condition of the entire fleet and that of a single aircraft, ignoring the potential shortage of maintenance resources, and the maintenance scheduling strategy is usually not optimal.

2. Due to the uncertainty of tasks and variety of aircrafts' health condition, maintenance strategy needs to be generated according to mission demands, aircrafts' health condition and resource limits. Those methods lack consideration into online decision making.

The fleet maintenance problem involves a lot of communication among aircrafts and maintenance teams, and Multi-Agent Modeling technique can imitate the communication and cooperation among agents to model complex systems (Budenske, Newhouse, Bonney & Wu, 2001), and has been successfully applied in many fields of manufacturing, especially dynamic and distributed scheduling problems. Through communication and cooperation can aircrafts and maintenance teams acquire the health condition of the whole fleet, and the working condition of maintenance teams. Meanwhile, the fleet maintenance problem is an N-P hard problem, and a common solution to N-P hard problems is heuristic searching. Heuristic rules can be integrated into agents to help overcome the N-P hardness, and is a guide to the intelligent allocation of maintenance tasks (Yang & Hu, 2007). In one word, Multi-Agent Modeling is suitable for solving the aircraft fleet maintenance problem.

This paper is the application of Multi-Agent System (*MAS*) to aircraft fleet maintenance scheduling. In this article, the idea of *MAS* and heuristic rules is adopted, and the dynamic intelligent maintenance decision making among an aircraft fleet with multiple maintenance teams is achieved to provide technical support for the online maintenance decision making. The purpose of this paper is to propose a multi-agent model, which can not only react to dynamic events, but can also generate schedules for maintenance jobs, to help design a fleet maintenance Decision Support System (DSS).

The remainder of this paper is organized as follows. Section 2 presents the description of the fleet maintenance problem. In Section 3, the *MAS* model for fleet maintenance scheduling is described, where the heuristic rules are put forward. The algorithm in which the dynamic problem is solved and schedules are generated is discussed in Section 4. Section 5 provides a case study of a mission oriented aircraft fleet to demonstrate the proposed method. Finally, concluding remarks and further study are provided in Section 6.

2. FLEET CBM PROBLEM DESCRIPTION

Consider an aircraft fleet containing m aircrafts and n maintenance teams ($n<m$) face continuous combat missions, in which a single mission requires l aircrafts (l is dynamic and $l \leq m$). Each aircraft contains p LRMs whose RUL can be estimated. All maintenance teams are of the same ability, namely the same LRM requires the same Mean Maintenance Time (MMT), while different LRMs require different MMTs. The basic assumptions of the problem are listed below.

1. The current mission is known, namely the upcoming mission and mission interval duration are known, while future missions are unknown.

2. Consider in-site maintenance only, so maintenance method is "replace and repair", and parts are repaired as good as new, namely the RUL of replaced LRMs reach the top.

3. The RUL of each LRM in each aircraft decreases with mission time, or RUL doesn't decrease without a mission. Moreover, due to the differences in historical missions, the initial RUL of different LRMs in different aircrafts are different.

4. Spare parts in each team are sufficient, namely spare parts are always available whenever a maintenance task is required.

5. The estimation of RUL is accurate, so the case in which wrong strategy led by wrong estimations won't occur.

6. Each team can work on only one aircraft at one time, and each aircraft can be repaired by only one team at one time.

After the whole fleet return from the previous mission, each aircraft checks its own health condition, estimating RULs and comparing the RULs with maintenance thresholds to decide a possible maintenance. There can be one or more threshold (Camci, Valentine & Navarra, 2007), and in this article two thresholds are required, namely the Required Maintenance Threshold τ and the Opportunistic Maintenance Threshold T. Those two thresholds divide the aircraft into three health states. When $RUL \leq \tau$ the state is identified as the required maintenance state S_3 and a maintenance is required immediately. When $RUL>T$, the state is identified as the no maintenance state S_1 and no maintenance is scheduled. When RUL is between these two thresholds $\tau < RUL \leq T$, the state is identified as the opportunistic maintenance state S_2 and a possible maintenance task depends on the states of other aircrafts and the occupation of maintenance teams. T & τ can be set according to mission or by experience. For instance, τ must exceed the time before the aircraft returns from the next mission.

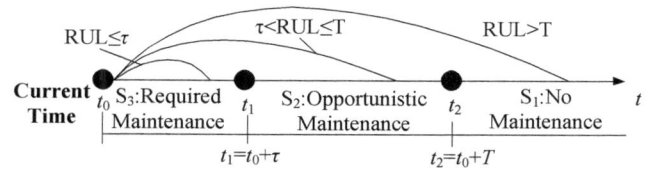

Figure 1. Maintenance thresholds and aircraft states

The objective of this problem is to maximize the availability of the fleet while the number of maintenance actions is satisfactory, and the basic constraints of the problem are:

1. The number of available aircrafts heading for the upcoming mission r must satisfy $r \geq l$.

2. The number of currently available teams s must satisfy $s \leq n$.

According to the description towards the problem above, when the fleet return from the previous mission, each aircraft checks its own health state S_i at the current time t_0, and reports to the maintenance center. The maintenance center verifies all the reports, organizes and coordinates maintenance tasks guided by a set of heuristic rules, and allocate maintenance tasks to suitable maintenance teams. Maintenance teams then execute maintenance tasks according to the maintenance center. When a maintenance task finishes, the fleet wait to execute the upcoming mission.

Each aircraft in the fleet will be repaired according to its condition. To all aircrafts, the combination of all maintenance decisions within the whole fleet forms a group of fleet CBM strategies aimed at utilizing the RULs of all aircrafts and the idle time of maintenance teams, in order to rationalize maintenance resources within the whole fleet.

3. THE FLEET CBM MODEL BASED ON MULTI-AGENT

The fleet CBM process involves a huge amount of communication among aircrafts, maintenance teams and the maintenance center. Moreover, maintenance teams and the maintenance center need to react to dynamic situations to make maintenance decisions and solve the problem, thus it can be regarded as a complex system (Zhang & Li, 2010), and one promising solution to complex systems is MAS. In MAS, an agent can be regarded as a self-directed software object with its own value system and a means to communicate with other agents (Baker, 1998), while the whole MAS can be regarded as "a loosely coupled network of problem solvers that work together to solve problems that are beyond the individual capabilities or knowledge of each problem solver" (Durfee, 1988). The fleet CBM process can be mapped into a similar MAS, where CBM strategies can be obtained via agents themselves and the communication between agents.

3.1. Model Framework

Through the analysis of the fleet *CBM* process, the physical entities can be abstracted into two types of agents, namely the Aircraft Agent (*AA*) and the Maintenance Agent (*MA*), and the dynamic process of management and coordination is abstracted into the Management and Coordination Agent (*MCA*).

AA is the abstract of an aircraft, it describes the inherent characteristics, the reliability characteristics, and is responsible for generating maintenance requirements. *MA* is the abstract of maintenance teams, and is responsible for specific maintenance process.

MCA is the abstract of the whole process of scheduling and intelligent allocating of maintenance tasks, it is driven by events, and is responsible for adjusting the whole process of maintenance, and obtaining the fleet maintenance strategy.

A 2-layer structure of *MAS* (Feng, Zeng & Kang, 2010) is applied to model the problem, each layer indicating the global scheduling and local scheduling, as shown in Figure 2.

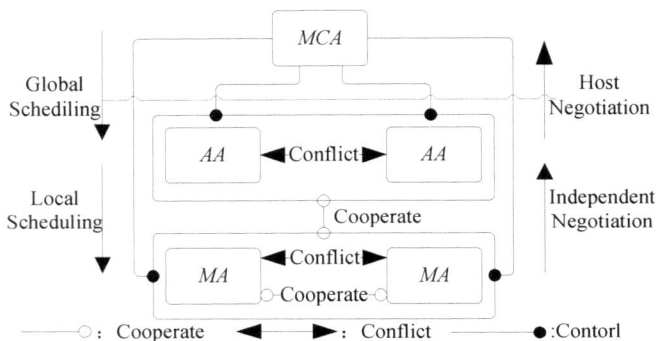

Figure 2. fleet *CBM MAS* model framework

Global Scheduling is conducted by *MCA*. When *MCA* receive the reports from *AA*s, it coordinates and controls the whole process and generates the overall maintenance strategy, to globally rationalize maintenance resources.

Local Scheduling is conducted between *AA*s and *MA*s, aimed at the negotiation in specific maintenance tasks.

3.2. Heuristic Rule-based Agent Negotiating Mechanism

The Contract Net Protocol (*CNP*) (Smith, 1980) is one of the most widely used agent negotiating mechanisms. Through imitating the "Calling-Bidding-Winning-Signing" process in economic behavior, *CNP* realizes the allocation, dynamic adjusting and converting of tasks among agents (Tang, Zhu, Li & Lei, 2010). Based on the *CNP*, the rationalization of the fleet *CBM* strategies is achieved.

In this article, all agents are assumed rational and friendly, their communication manifest cooperation and conflicts, which means that an agent is willing to cooperate with other agents, and maximize its own profit if possible. That

assumption caters for practical situations. For instance, each aircraft wishes to be repaired as early as possible. A maintenance team needs cooperation to repair all aircrafts, but wishes to repair as many aircrafts as possible.

Since the *MAS* model applies the 2-layer structure, the negotiating between agents is also divided into two layers, namely the Host Negotiating and the Independent Negotiating. As proved above, the problem of a fleet maintenance with multiple maintenance teams is N-P hard, it's difficult to obtain the satisfying solution. So in each layer, negotiation must follow its corresponding heuristic rules, as described below.

3.2.1. Heuristic Rules in Independent Negotiation

In Independent Negotiation, idle *MA*s communicate with *AA*s to obtain local maintenance strategies, the alternative maintenance decision making heuristic rules are listed below.

1. Aircrafts in the required maintenance state S_3

- The shortest total waiting time principle: all aircrafts in the required maintenance state S_3 are scheduled to shorten the average waiting time, or to even the working time of all maintenance teams. This rule is marked "Rule 11a".

- The most repairs within limited interval principle: once a maintenance team is idle, a maintenance task is performed on the aircraft with the shortest *MMT*. This rule is marked "Rule 11b".

- Single team with widest repair time margin principle: as many aircrafts are repaired by as few maintenance teams as possible, so as to leave the most teams idle, in case unexpected failures occur. This rule is marked "Rule 11c".

2. Aircrafts in the opportunistic maintenance state S_2

- The most repairs within limited interval principle: once a maintenance team is idle, a maintenance task is performed on the aircraft with the shortest *MMT*. This rule is marked "Rule 12a".

3.2.2. Heuristic Rules in Host Negotiation

In Host Negotiation, the *MCA* communicates with *AA*s to obtain global maintenance strategies, generates a group of local maintenance tasks and dispatches tasks to corresponding *MA*s. The whole process is listed below:

Assume that the number of aircrafts needed for the upcoming mission is l_n. *AA*s first report their health states S_t to the *MCA*. The *MCA* analyses all data reported and confirms the number of *AA*s in the required maintenance state S_3 m_3, the number of *AA*s in the opportunistic maintenance state S_2 m_2, and the number of *AA*s in the no maintenance state S_1 m_1. The *MCA* then calculates the number of repairable aircrafts within the interval m_4 according to Rule 11a, Rule 11b and Rule 11c respectively, and gets the maximum number m_4, and the optimal rule is expressed as $Pro(Rul_i)$. The number of combat-ready AAs $m_a = m_1 + m_2 + m_4$. The alternative maintenance decision making heuristic rules are shown in Figure 3.

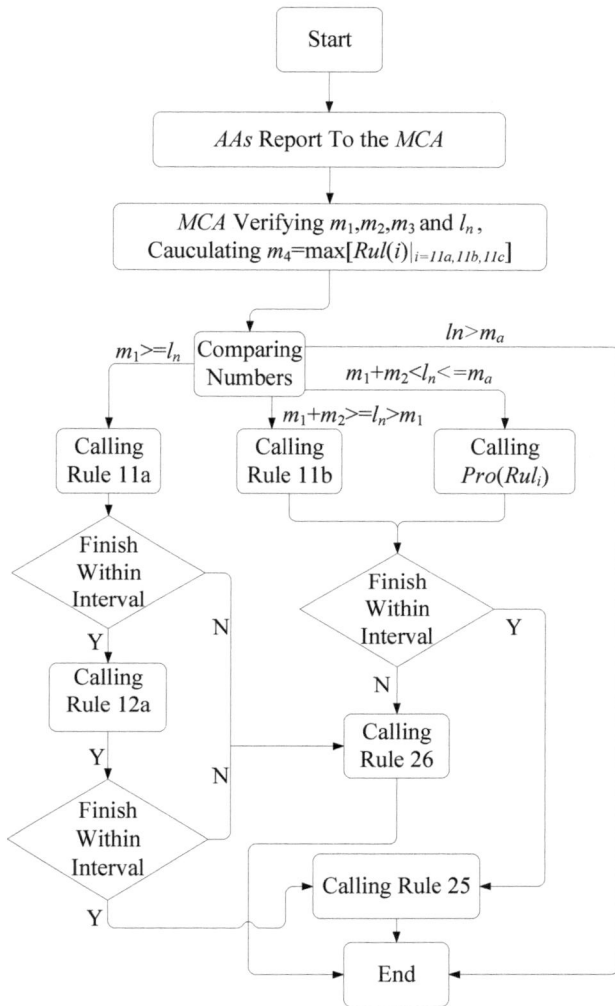

Figure 3. The flow chart of generating maintenance tasks

Based on the analysis above, the alternative maintenance decision making heuristic rules are listed below.

1. If $m_a < l_n$, then mission fails. This rule is marked "Rule 21".

2. If the number of AAs in the no maintenance state S_1 satisfies $m_1 \geq l_n$, then AAs in the no maintenance state S_1 are put on mission first, and AAs in the required maintenance state S_3 are repaired according to Rule 11a. When current task finishes, AAs in the opportunistic maintenance state S_2 are repaired according to Rule 12a, where AAs with the shortest MMT are repaired with high priority. This rule is marked "Rule 22".

3. If $m_1 < l_n \leq m_1 + m_2$, then AAs in the required maintenance state S_3 are repaired according to Rule 11b. This rule is marked "Rule 23".

4. If $m_1 + m_2 < l_n \leq m_1 + m_2 + m_4$, then AAs in the no maintenance state S_1 and the opportunistic maintenance state S_2 are put on mission first, and AAs in the required maintenance state S_3 are repaired according to $Pro(Rul_i)$. This rule is marked "Rule 24".

5. When the interval ends, each aircraft checks its health state again, and reports to the MCA. Then the MCA analyses the reported data and select l_n AAs with the shortest RUL out of all combat-ready AAs (AAs in the opportunistic maintenance state S_2, AAs in the no maintenance state S_1 and repaired AAs) to execute the mission. This rule is marked "Rule 25".

6. When mission starts, if there exists still AAs in the opportunistic maintenance state S_2 required maintenance state S_3 among all the left-over AAs, then those AAs are repaired according to Rule 11a and Rule 12a respectively. This rule is marked "Rule 26".

3.2.3. Agent Behavior in fleet CBM

Based on the analysis of the process of fleet CBM, the MAS model framework and the heuristic rules on solving maintenance strategies, the Agent Ability Chart (Feng, 2009) in fleet CBM is established, which finally defines agents' attributes and behaviors of function & fault, laying the foundation of solving the maintenance strategies.

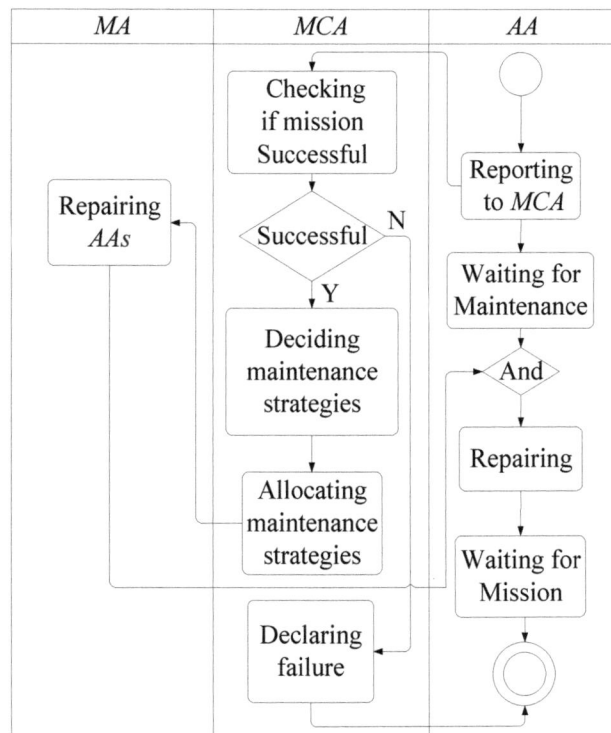

Figure 4. The Agent ability chart in fleet CBM

4. MAINTENANCE STRATEGY SOLVING ALGORITHM

AUML (Bauer, Müller & Odel, 2001), based on object oriented design, is a typical agent oriented modeling technique. It provides a uniform agent oriented modeling mechanism, and doesn't restrict too much on modeling process. The process of fleet *CBM* can be realized through the communication between *AA*s and *MA*s. With the help of AUML, the negotiating models between agents are established based on *CNP*.

Since the *CBM* model involves communication between and within layers, the problem is relatively complex. As space is limited, three of the most typical maintenance schemes are illustrated. These three corresponding algorithms are listed below.

4.1. The Shortest Total Maintenance Waiting Time Maintenance Scheme Negotiating Algorithm

This scheme is relatively integrated, which involves cooperative and competitive negotiations. The algorithm is listed below.

4.1.1. Cooperative Negotiation

Cooperative negotiation is required before a maintenance task starts. It's aimed at calculating the whole maintenance time needed and allocating each *MA* its corresponding maintenance time.

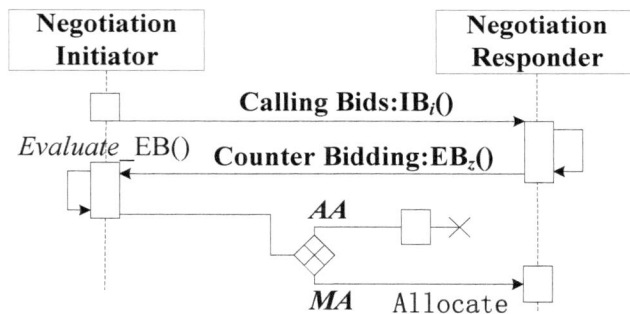

Figure 5. The cooperative negotiation mechanism

Step 1: The negotiation initiator calling for bids.

The first idle *MA* i (a random *MA* if there exists more than one) calls other *MA*s and all *AA*s for bids $IB_i(t_i|ta_i,tb_i)$, where t_i represents the latest bid time allowed, ta_i represents the earliest idle time of other *MA*s (Time to finish current task), tb_i represents the maintenance duration needed.

Step 2: The negotiation responders counter-bidding.

*MA*s and *AA*s assess their own status and counter-bid before t_i. The counter-bids from *MA*s are represented as $EB_j(t_j|ta_j)$, where t_j represents the waiting time, ta_j

represents the earliest idle time. While the counter-bids from *AA*s are represented as $EB_k(t_k|tb_k)$, where t_k represents the waiting time, ta_k represents the maintenance time needed.

Step 3: The negotiation initiator responding to all counter-bids

The negotiation initiating *MA* counts all counter-bids. Assume that m is the number of counter-bids from *MA*s and n is the number of counter-bids from *AA*s. Then based on the Shortest Total Maintenance Waiting Time Principle, The negotiation initiating *MA* calculates the Allocated Maintenance Time (*AMT*) to *MA* j through function *Evaluate_EB*(),

$$AMT_j = (\sum_1^m ta_j + \sum_1^n tb_k)\ /\ m - ta_j \qquad (1)$$

and responds to each *MA* its *AMT*.

4.1.2. Competitive Negotiation

Competitive negotiation is required during the process of specific maintenance tasks. It's aimed at confirming maintenance strategy and realizing maintenance tasks.

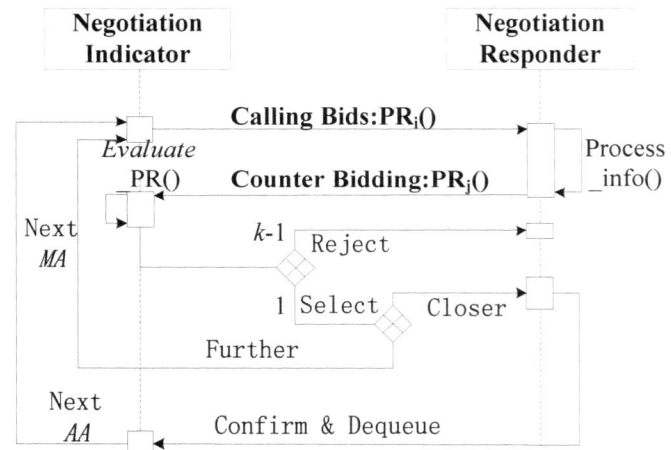

Figure 6. The competitive negotiation mechanism

Step 1: *MA* calling for bids.

The first idle *MA* i (a random *MA* if there exists more than one) calls all *AA*s for bids $PR_i(T_i|AMT_i)$, where T_i represents the latest bid time allowed, AMT_i represents the allocated maintenance time.

Step 2: *AA*s counter-bidding.

*AA*s assess their own status through function *Process_info*(). If it's within the candidate queue, then counter-bid before t_i. The counter-bids from *AA*s are represented as $PR_j(T_j|MMT_j)$, where T_j represents the waiting time, MMT_j represents the maintenance time needed.

Step 3: *MA* assessing all counter-bids

MA counts all counter-bids and assesses them through function *Evaluate_EB*(), ranking all counter-bidding *AA*s according to the length of *MMT* and selecting the candidate *a* with the closest *MMT* to *AMT*.

Step 4: *MA* judging whether to stop bidding.

MA updates its *AMT*: $AMT_{temp}=AMT-MMT_a$ for the moment. If abs (AMT_{temp}) < abs (AMT), then *MA* updates the $AMT=AMT_{temp}$ and responds to the selected *AA* and the selected *AA* then dequeues, repeat Step1 ~ Step3. Else, *MA* stops the current process of bidding and starts repairing all selected *AA*s.

Step 5: Other *MA*s start bidding according to the idle time order (a random *MA* if there exists more than one), repeat Step1 ~ Step4.

4.2. The Most Repairs Within the Limited Interval Maintenance Scheme Negotiating Algorithm

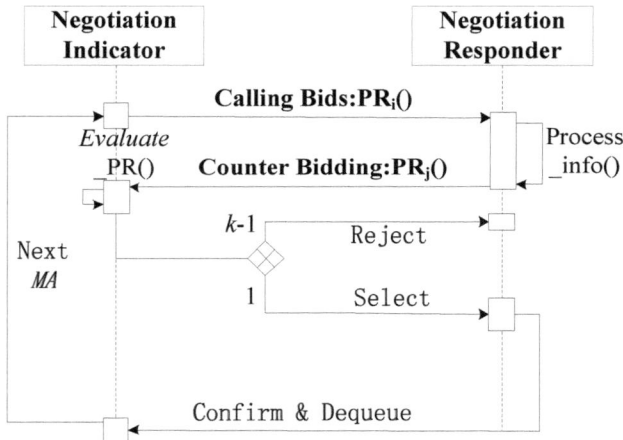

Figure 7. The most repairs within the limited interval maintenance scheme negotiation mechanism

Step 1: *MA* calling for bids.

The first idle *MA i* (a random *MA* if there exists more than one) calls all *AA*s for bids $PR_i(T_i|LMT_i)$, where T_i represents the latest bid time allowed, LMT_i represents the longest maintenance time.

Step 2: *AA*s counter-bidding.

*AA*s assess their own status through function *Process_info*(). If it's within the candidate queue, then counter-bid before t_i. The counter-bids from *AA*s are represented as $PR_j(T_j|MMT_j)$, where T_j represents the waiting time, MMT_j represents the maintenance time needed.

Step 3: *MA* assessing all counter-bids

MA counts all counter-bids and assesses them through function *Evaluate_EB*(), ranking all counter-bidding *AA*s according to the length of *MMT* and selecting the candidate *a* with the shortest *MMT*. Then repair task starts, when task finishes, *MA* updates its $LMT=LMT-MMT_a$.

Step 4: The repaired *AA* then dequeues. Other *MA*s start bidding according to the idle time order (a random *MA* if there exists more than one), repeat Step1 ~ Step3.

4.3. Single Team with Widest Repair Time Margin Maintenance Scheme Negotiating Algorithm

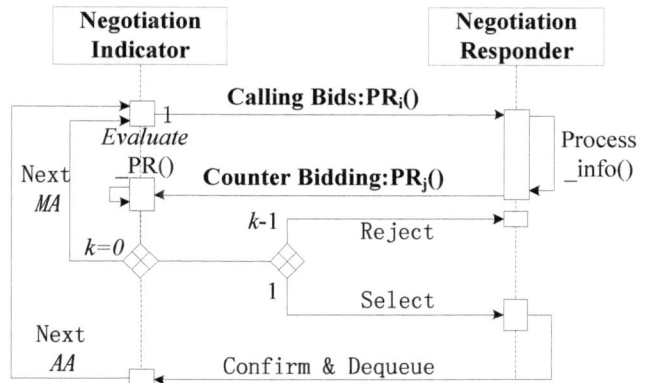

Figure 8. Single team with widest repair time margin maintenance scheme negotiation mechanism

Step 1: *MA* calling for bids.

The first idle *MA i* (a random *MA* if there exists more than one) calls all *AA*s for bids $PR_i(T_i|LMT_i)$, where T_i represents the latest bid time allowed, LMT_i represents the longest maintenance time.

Step 2: *AA*s counter-bidding.

*AA*s assess their own status through function *Process_info*(). If it's within the candidate queue, then counter-bid before t_i. The counter-bids from *AA*s are represented as $PR_j(T_j|MMT_j)$, where T_j represents the waiting time, MMT_j represents the maintenance time needed.

Step 3: *MA* assessing all counter-bids

MA counts all counter-bids and assesses them through function *Evaluate_EB*(), ranking all counter-bidding *AA*s according to the length of *MMT* and selecting the candidate *a* with the closest *MMT* to *LMT*. Then the selected *AA* dequeues.

Step 4: *MA* updates its $LMT=LMT-MMT_a$ and repeats Step1 ~ Step3, till there's no suitable candidate. Then stop bidding and start repairing all selected *AA*s.

Step 5: Other *MA*s start bidding according to the idle time order (a random *MA* if there exists more than one), repeat Step1 ~ Step4.

5. CASE STUDY

A typical continuous mission of a fleet is presented to verify the proposed fleet *CBM* decision making strategy. Assume a fleet consisting of 10 aircrafts, each monitoring the condition of two *LRM*s and predicting their corresponding *RUL*s, which carries on a 3-wave mission. The time property of the mission is listed in Table 1

No.	Start Time (h)	Mission (h)	Interval (h)
1	1	2	2
2	5	3	1
3	9	2.5	-

Table 1. Time property of the mission

During the mission, there exist two maintenance teams to support the whole fleet. The maintainability (*MMT*) of the two *LRM*s are listed in Table 2

LRM Sorts	MMT (h)
LRM_1	0.5
LRM_2	1

Table 2. Reliability and maintainability data

To effectively verify this method, assume that some aircrafts are in the no maintenance state S_1 while others are in the opportunistic maintenance state S_2, hence all aircrafts can take part in the first mission, and no maintenance is considered before the first wave. The initial *RUL* of all aircrafts in the fleet are listed in Table 3

LRM \ Aircraft	1	2	3	4	5
LRM_1	16.3	8.5	12.6	5.6	19.1
LRM_2	18.1	18.2	5.9	10.9	19.2

LRM \ Aircraft	6	7	8	9	10
LRM_1	5.1	19.1	16.0	8.4	15.8
LRM_2	19.4	9.7	5.8	18.3	19.1

Table 3. The initial *RUL* of the fleet

Since the future missions are unknown, the maintenance thresholds can be decided as: τ is the time before the aircraft return from the next mission, and T is $T = 2 \times \tau$. On each inspection, the maintenance thresholds are listed in Table 4

Maintenance thresholds(h)	First Inspection	Second Inspection
τ	3	2.5
T	6	5

Table 4. Maintenance thresholds on each inspection

Assume that the first wave requires 8 aircrafts, the second wave requires 6 and the third wave requires all aircrafts. Based on our former assumptions and the maintenance decision making rules, maintenance strategies can be obtained as listed in Table 5, where the 2nd and 3rd column indicates the number of aircrafts repaired in the team. For instance, "3,4" means aircraft 3 and 4 are repaired in team 1. The 4th column indicates the spared aircrafts from the mission, for instance, "1,5" indicates aircraft 1 and 5 are spared from this mission. The 5th column indicates whether this mission is successful.

Waves	Queue 1	Queue 2	Spared	Mission Succeed
1	null	null	1,5	Y
2	3,4	6,8	1,5,6,8	Y
3	2,9	7	null	Y

Table 5. Fleet maintenance strategies.

Traditional *CBM* methods, which concentrates more on "timely" maintenance decision making rather than "online", can hardly make maintenance decisions online, so is not comparable with the *MAS* method. To make the comparison possible, *MAS* is applied to model traditional *CBM* policy, which assumes that an aircraft is repaired only when it comes to the required maintenance state S_3, relies on a single threshold, and ignores the states of the whole fleet and the maintainability of limited teams. Assume that the initial state, mission time property and maintainability of teams are the same, and the fleet maintenance strategies are listed in Table 6

Waves	Queue 1	Queue 2	Spared	Mission Succeed
1	null	null	1,5	Y
2	null	null	1,5,6,10	Y
3	3	4,6	8	N

Table 6. Single threshold fleet maintenance strategies

The table shows that before the 3rd wave, aircraft 3,4,6,8 all need repairing, and the total time required is $3h$, which exceeds the maximum time teams can offer, so mission fails.

The case above shows that the 2-thresholds *CBM* policy is superior to traditional single-threshold *CBM* policy in both flexibility and results.

INTERNATIONAL JOURNAL OF PROGNOSTICS AND HEALTH MANAGEMENT, VOL.3 (2012)

6. CONCLUSION

In this paper, a fleet *CBM* intelligent decision making method based on *MAS* and heuristic rules is proposed, which is a technical support for fleet online maintenance decision making, and can help design a fleet maintenance Decision Support System (DSS). A fleet consisting of 10 aircrafts and 2 teams is illustrated to verify the correctness and feasibility of this method.

To avoid the local optimal solution, host negotiating is proposed to coordinate the global maintenance strategies, which can not only guarantee the correctness and feasibility of the solution, but also optimize the global maintenance strategy.

A 2-thresholds *CBM* policy is proposed, and results show that the 2-thresholds *CBM* policy is superior to traditional single-threshold *CBM* policy in both flexibility and results, while the requirement to decide maintenance threshold is much higher.

This method mainly concentrates on the strategy itself. With suitable improvement, this method can be modified to optimize maintenance resources.

This method is based on an assumption that the *RUL* estimation is accurate, and the maintenance strategies are based on accurate *RUL*s. Considering the defects in failure prognostics technology, further study needs to discuss the relationship between the accuracy of the *RUL* estimates and the availability of the fleet, where *PHM* uncertainty management will be considered.

ACKNOWLEDGEMENT

This work is partially supported by the Fundamental Research Funds for the Central Universities of China (No. YWF-12-LSJC-001).

NOMENCLATURE

AA	Aircraft Agent
AMT	Allocated Maintenance Time
CBM	Condition-Based Maintenance
CNP	Contract Net Protocol
LMT	Longest Maintenance Time
LRM	Line Replaceable Module
MA	Maintenance Agent
MAS	Multi-Agent System
MCA	Management and Coordination Agent
MMT	Mean Maintenance Time
PHM	Prognostics and Health Management
RUL	Remaining Useful Life

REFERENCES

Jiang, R. Y. & Murthy, D. N. P. (2008). *Maintenance: Decision Models for Management*. Beijing: Science Press.

Bengtsson, M. (2004). Condition Based Maintenance System Technology - Where is Development Heading. *Congress Report on the 17th Euromaintenance*, May, Barcelona, Spain. http://www.idp.mdh.se/forskning/amnen/produktproces s/projekt/cbm/publi-kationer/em04%20paper%20Marcus%20Bengtsson.pdf.

Sun, B., Zeng, S.K., Kang, R. & Pecht, M. G. (2012). Benefits and Challenges of System Prognostics. *Reliability. IEEE Transactions on*, vol. 61(2), pp. 323-335. doi:10.1109/TR.2012.2194173

Goebel, K.F., Saha B., Saxena, A., Celaya, J.R. & Christophersen, J.P. (2008). Prognostics in Battery Health Management. *Instrumentation & Measurement Magazine, IEEE*, vol. 11(4), pp. 33-40. doi:10.1109/MIM.2008.4579269

Wen, Z.H. & Liu, Y.P. (2011). Applications of Prognostics and Health Management in aviation industry. *Prognostics and System Health Management Conference*. May 24-25, Shenzhen, China. doi:10.1109/PHM.2011.5939539

Barata, J., Guedes, S.C., Marseguerra, M. & Zio, E. (2002). Simulation modeling of repairable multi-component deteriorating systems for 'on condition' maintenance optimization. *Reliability Engineering & System Safety*, vol. 76(3), pp. 255–264. doi:10.1016/S0951-8320(02)00017-0

Jardine, A.K.S., Lin, D. & Banjevic, D. (2006). A review on machinery diagnostics and prognostics implementing condition-based maintenance. *Mechanical Systems and Signal Processing*, vol. 20(7), pp. 1483-1510. doi:10.1016/j.ymssp.2005.09.012

Garey, M.R. & Johnson, D.S. (1979). *Computers and Intractability: A Guide to the Theory of NP-Completeness*. New York, USA: WH Freeman & Co.

Doganay, K. & Bohlin, M. (2010). Maintenance plan optimization for a train fleet. *WIT Transactions on The Built Environment*, vol. 114(12), pp. 349-358. doi:10.2495/CR100331

Bai, F. (2009). *Methods of Scheduling and Condition Based Mainteance Decision Making in Civil Aero Engine Fleet*. Doctoral dissertation. Nanjing University of Aeronautics and Astronautics, Nanjing, China

Reimann, J., Kacprzynski, G., Cabral, D. & Marini, R. (2009). Using Condition Based Maintenance to Improve the Profitability of Performance Based Logistic Contracts. *Annual Conference of the Prognostics and Health Management Society*, Sep 27-Oct 1, San Diego, California. http://www.phmsociety.org/sites/phmsociety.org/files/p hm_submission/2009/phmc_09_52.pdf

Bivona, E. & Montemaggiore, G.B. (2005). Evaluating Fleet and Maintenance Management Strategies through

System Dynamics Model in a City Bus Company. *The 23rd International Conference of the System Dynamics Society*, July 17-21, Boston. http://www.systemdynamics.org/conferences/2005/proceed/papers/MONTE431.pdf

Dupuy, M.J., Wesely, D.E. & Jenkins, C.S. (2011). Airline fleet maintenance: Trade-off analysis of alternate aircraft maintenance approaches. *Systems and Information Engineering Design Symposium (SIEDS)*. April 29, Charlottesville, VA. doi:10.1109/SIEDS.2011.5876850

Cycon, J.P. (2011). The journey to incorporate health monitoring and condition based maintenance of Sikorsky commercial helicopters. *8th International Workshop on Structural Health Monitoring 2011: Condition-Based Maintenance and Intelligent Structures*, September 13-15, Stanford, CA. http://www.destechpub.com/links/catalogs/bookstore/advanced-materials-sciencetechnology/structural-health-monitoring/structural-health-monitoring-2011/#table

Zhou, R., Fox, B., Lee, H.P. & Nee., A.Y.C. (2004). Bus maintenance scheduling using multi-agent systems. *Engineering Applications of Artificial Intelligence*, vol. 17(6), pp. 623–630. doi:10.1016/j.engappai.2004.08.007

Papakostas, N., Papachatzakis, P., Xanthakis, V., Mourtzis, D. & Chryssolouris, G. (2010). An approach to operational aircraft maintenance planning. *Decision Support Systems*, vol. 48(4), pp. 604–612. doi:10.1016/j.dss.2009.11.

Budenske J., Newhouse J., Bonney J. & Wu J. (2001). Agent-based schedule validation and verification. *Systems, Man, and Cybernetics, 2001 IEEE International Conference on*. Oct 7-10, Tucson, AZ. doi:10.1109/ICSMC.2001.969920

Yang, S.L. & Hu, X.J. (2007). *Complex decision task modeling and solving methods*. Beijing: Science Press

Camci, F., Valentine, G.S. & Navarra, K. (2007). Methodologies for Integration of PHM Systems with Maintenance Data. *Aerospace Conference, 2007 IEEE*, March 3-10, Big Sky, MT. doi:10.1109/AERO.2007.352917

Zhang, G.J. & Li, Y.D. (2010). Agent-based modeling and simulation for open complex systems. *Informatics in Control, Automation and Robotics (CAR), 2010 2nd International Asia Conference on*, March 6-7, Wuhan. doi: 10.1109/CAR.2010.5456783

Baker, A.D. (1998). A survey of factory control algorithms that can be implemented in a multi-agent heterarchy: dispatching, scheduling and pull. *Journal of Manufacturing Systems*, vol. 17(4), pp. 297–320. doi: 10.1016/S0278-6125(98)80077-0

Durfee, E.H. (1988). *Coordination of Distributed Problem Solvers*. Boston, MA: Kluwer Academic Publishers.

Feng, Q, Zeng, S.K. & Kang, R. (2010) Multiagent-based modeling method for integrated logistic support of the carrier aircraft. *Systems Engineering and Electronics*, vol. 32(1), pp. 211-217.

Smith, R.G. (1980). The Contract Net Protocol: High-Level Communication and Control in a Distributed Problem Solver. *Computers, IEEE Transactions on*, vol. C-29(12), pp. 1104-1113. doi: 10.1109/TC.1980.1675516

Tang, S.Y., Zhu, Y.F., Li, Q. & Lei, Y.L. (2010). Survey of task allocation in multi Agent systems. *Systems Engineering and Electronics*, vol. 32(10), pp. 2155-2161. doi:0.3969/j/issn/1001-506X.2010.10.30

Feng, Q. (2009). *Mesuring and Modeling and Optimization Method of Effectiveness for Complex Materiel System based on Multi-Agent*. Doctoral dissertation. Beihang University, Beijing, China

Bauer, B., Müller, J.P. & Odell, J. (2001). Agent UML: a formalism for specifying multiagent software systems. *International Journal of Software Engineering and Knowledge Engineering*, vol. 11(3), pp. 1-12. doi:10.1142/S0218194001000517

BIOGRAPHIES

Dr. Qiang Feng is a reliability engineer and a member of the faculty of systems engineering at the School of Reliability and Systems Engineering at Beihang University in Beijing, China. He received his Ph.D. degree in reliability engineering and systems engineering from Beihang University and a B.S. degree in mechanical engineering from Beihang University. He has won a 1st and a 3rd prize for National Defense Science and Technology Progress Award. His current research interests include reliability engineering, reliability of complex product and integrated design of product reliability and performance.

Songjie Li is a graduate student at the School of Reliability and Systems Engineering at Beihang University in Beijing, China. He received a B.S. degree in mathematics and control theory from Beihang University and is a master candidate. He has won a 2nd prize in Graduate Scholarship twice in a row. His current research interests include reliability engineering, reliability and performance of complex product and system modeling & simulation.

Dr. Bo Sun is a reliability engineer and a member of the faculty of systems engineering at the School of Reliability and Systems Engineering at Beihang University in Beijing, China. He received his Ph.D. degree in reliability engineering and systems engineering from Beihang University and a B.S. degree in

mechanical engineering from the Beijing Institute of Mechanical Industry. His current research interests include prognostics and health management, physics of failure, reliability of electronics, reliability engineering, and integrated design of product reliability and performance. He has won a 1st and a 3rd prize for National Defense Science and Technology Progress Award. He has published over 40 papers and 2 book chapters (Reliability Design and Analysis, and Diagnostics, Prognostics, and System's Health Management). He is now a member of the Editorial Board

Author Index

Author Guidelines

The International Journal of Prognostics and Health Management (IJPHM) publishes scientific papers dealing with all aspects of prognostics, diagnostics, and system health management of complex engineered systems. High quality articles focused on assessing the current status and predicting the future condition of an engineered component and/or system of components. Such articles may come from a variety of disciplines, including electrical, electronics, mechanical, civil, and chemical engineering, computer and materials science, reliability, test and measurement, artificial intelligence, physics, and economics.

Copyright

The Prognostic and Health Management Society advocates open-access to scientific data and uses a Creative Commons license for publishing and distributing any papers. A Creative Commons license does not relinquish the author's copyright; rather it allows them to share some of their rights with any member of the public under certain conditions whilst enjoying full legal protection. By submitting an article to the International Conference of the Prognostics and Health Management Society, the authors agree to be bound by the associated terms and conditions including the following: As the author, you retain the copyright to your Work. By submitting your Work, you are granting anybody the right to copy, distribute and transmit your Work and to adapt your Work with proper attribution under the terms of the Creative Commons Attribution 3.0 United States license. You assign rights to the Prognostics and Health Management Society to publish and disseminate your Work through electronic and print media if it is accepted for publication. A license note citing the Creative Commons Attribution 3.0 United States License, as shown below, needs to be placed in the footnote on the first page of the article.

First Author et al. This is an open-access article distributed under the terms of the Creative Commons Attribution 3.0 United States License, which permits unrestricted use, distribution, and reproduction in any medium, provided the original author and source are credited.

Ethics

Contributions to IJPHM must report original research and will be subjected to review by referees at the discretion of the Editor. IJPHM considers only manuscripts that have not been published elsewhere (including at conferences), and that are not under consideration for publication or in press elsewhere. Moreover, it is the responsibility of the author to ensure that any data or information submitted complies with the export-control regulations of the author's home country (e.g., International Traffic in Arms Regulations (ITAR) in the United States). IJPHM honors code of conduct provided by the Committee of Publication Ethics (COPE). More details on IJPHM policies and publication ethics can be found online.

Submission Types

IJPHM publishes full-length regular papers, technical briefs, communications, and survey papers.

Full-Length Regular Papers should describe new and carefully confirmed findings, and experimental procedures and results should be given in detail sufficient for others to replicate the work. A full paper should be long enough to describe and interpret the work clearly, placing it in the context of other research.

Technical Briefs usually describe a single result, experiment, or technique of general interest for which a short treatment is appropriate. A short paper should be long enough to describe experimental procedures and clearly, and interpret the results in the context of other research.

Communications are a separate class of short manuscripts that are subject to an expedited review process. Appropriate items include (but are not limited to) rebuttals and/or counterexamples of previously published papers. A short communication is suitable for recording the results of complete small investigations or giving details of new models or hypotheses, innovative methods, techniques or apparatus. The style of main sections need not conform to that of full-length papers. Short communications are 2 to 4 printed pages in length. The Editors will review these submissions internally, and request outside review when appropriate.

Survey Papers covering emerging research topics in PHM are also published, and unsolicited manuscripts of a tutorial or review nature are welcome. However, prospective authors of survey papers should contact in advance the Editor-in-Chief in order to assess the possible interest of the topic to IJPHM. Papers describing specific current applications are encouraged, provided that the designs represent the best current practice, detailed characteristics and performance are included, and they are of general interest.

Prospective authors should note that for any type of IJPHM content, poorly documented papers using "proprietary" techniques will be rejected. Moreover, excessive "branding" within a paper also cause for rejection; e.g., "The team used the magical CompanyBrand™ preprocessing to prepare the data to extract the amazing CompanyBrand™-proprietary features (which we can't tell you about)." Papers should present techniques and results clearly and objectively.

Although bound editions will be available for purchase, IJPHM is fundamentally an online journal. As such, we are able to have a very fast turnaround time. We will acknowledge receipt of submissions within three business days, and we intend to rigorously review and return a decision to the authors in approximately 8-12 weeks. Thus, papers may be published in a very short time, allowing your research to be available to the scientific community when it is most relevant.

Option to Present Your Work at a Conference

PHM Society publications have maintained high quality standards for both its Conferences and the Journal. Highest quality conference papers are also invited to be published in the Journal. However, since 2012 IJPHM provides an option to the journal authors to present their journal paper at one of the upcoming PHM conferences.

Authors are reminded that the paper must be journal quality and adhering to the journal template. The paper will be reviewed as per journal review standards and if accepted a presentation slot will be reserved at the target conference. The paper will be published in the journal archives and linked through conference proceedings.

Benefits
- A journal publication of your high quality research work
- A peer review of your work by experts in the field
- A chance to present your work to the targeted audience
- No reworking required to publish in the Journal
- A shortened review cycle to journal publishing

Risks
- Rejected papers will not automatically be considered for the conference and may additionally miss the submission deadline.
- If re-submitted for the conference, they will be reviewed subject to conference review criteria

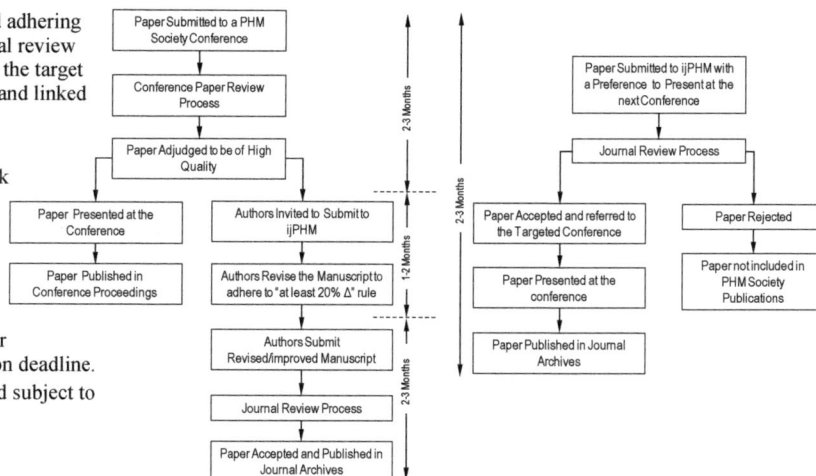

Flowchart (left path):
Paper Submitted to a PHM Society Conference → Conference Paper Review Process → Paper Adjudged to be of High Quality → Paper Presented at the Conference → Paper Published in Conference Proceedings → Authors Invited to Submit to IJPHM → Authors Revise the Manuscript to adhere to "at least 20% Δ" rule → Authors Submit Revised/Improved Manuscript → Journal Review Process → Paper Accepted and Published in Journal Archives (2-3 Months)

Flowchart (right path):
Paper Submitted to IJPHM with a Preference to Present at the next Conference → Journal Review Process → Paper Accepted and referred to the Targeted Conference → Paper Presented at the conference → Paper Published in Journal Archives / Paper Rejected → Paper not included in PHM Society Publications (2-3 Months)

www.ingramcontent.com/pod-product-compliance
Lightning Source LLC
Chambersburg PA
CBHW041718210326
41598CB00007B/692